HEART FUNCTION IN HEALTH AND DISEASE

DEVELOPMENTS IN CARDIOVASCULAR MEDICINE

121. S. Sideman, R. Beyar and A.G. Kleber (eds.): Cardiac Electrophysiology, Circulation, and Transport. Proceedings of the 7th Henry Goldberg Workshop (Berne, Switzerland, 1990). 1991. ISBN 0–7923–1145–0.
122. D.M. Bers: Excitation-Contraction Coupling and Cardiac Contractile Force. 1991. ISBN 0–7923–1186–8.
123. A.-M. Salmasi and A.N. Nicolaides (eds.): Occult Atherosclerotic Disease. Diagnosis, Assessment and Management. 1991. ISBN 0–7923–1188–4.
124. J.A.E. Spaan: Coronary Blood Flow. Mechanics, Distribution, and Control. 1991. ISBN 0–7923–1210–4.
125. R.W. Stout (ed.): Diabetes and Atherosclerosis. 1991. ISBN 0–7923–1310–0.
126. A.G. Herman (ed.): Antithrombotics. Pathophysiological Rationale for Pharmacological Interventions. 1991. ISBN 0–7923–1413–1.
127. N.H.J. Pijls: Maximal Myocardial Perfusion as a Measure of the Functional Significance of Coronary Arteriogram. From a Pathoanatomic to a Pathophysiologic Interpretation of the Coronary Arteriogram. 1991. ISBN 0–7923–1430–1.
128. J.H.C. Reiber and E.E. v.d. Wall (eds.): Cardiovascular Nuclear Medicine and MRI. Quantitation and Clinical Applications. 1992. ISBN 0–7923–1467–0.
129. E. Andries, P. Brugada and R. Stroobrandt (eds.): How to Face 'the Faces' of Cardiac Pacing. 1992. ISBN 0–7923–1528–6.
130. M. Nagano, S. Mochizuki and N.S. Dhalla (eds.): Cardiovascular Disease in Diabetes. 1992. ISBN 0–7923–1554–5.
131. P.W. Serruys, B.H. Strauss and S.B. King III (eds): Restenosis after Intervention with New Mechanical Devices. 1992. ISBN 0–7923–1555–3.
132. P.J. Winter (ed.): Quality of Life after Open Heart Surgery. 1992. ISBN 0–7923–1580–4.
133. E.E. van der Wall, H. Sochot, A. Righetti and M.G. Niemeyer (eds.): What is new in Cardiac Imaging? SPECT, PET and MRI. 1992. ISBN 0–7923–1615–0.
134. P. Hanrath, R. Uebis and W. Krebs (eds.); Cardiovascular Imaging by Ultrasound. 1992. ISBN 0–7923–1755–6.
135. F.H. Messerli (ed.): Cardiovascular Disease in the Elderly, 3rd ed. 1992. ISBN 0–7923–1859–5.
136. J. Hess and G.R. Sutherland (eds.): Congential Heart Disease in Adolescents and Adults. 1992. ISBN 0–7923–1862–5.
137. J.H.C. Reiber and P.W. Serruys (eds.): Advances in Quantitative Coronary Arteriography. 1992. ISBN 0–7923–1863–3.
138. A.-M. Salmasi and A.S. Iskandrian (eds): Cardiac Output and Regional Flow in Health and Disease. (in prep.). ISBN 0–7923–1911–7.
139. J.H. Kingma, N.M. van Hemel and K.I. Lie (eds.): Atrial Fibrillation, a Treatable Disease? 1992. ISBN 0–7923–2008–5.

HEART FUNCTION IN HEALTH AND DISEASE

Proceedings of the Cardiovascular Program sponsored by the Council of Cardiac Metabolism of the International Society and Federation of Cardiology during the Regional Meeting of the International Union of Physiological Sciences, Prague, Czechoslovakia, June 30 - July 5, 1991.

edited by

Bohuslav Ostadal
Director, Institute of Physiology
Czechoslovak Academy of Sciences
Prague, Czechoslovakia

Naranjan S. Dhalla
Distinguished Professor and Head
Division of Cardiovascular Sciences
St. Boniface General Hospital Research Centre
University of Manitoba
Winnipeg, Canada

Kluwer Academic Publishers
Boston/Dordrecht/London

Distributors for North America:
Kluwer Academic Publishers
101 Philip Drive
Assinippi Park
Norwell, Massachusetts 02061 USA

Distributors for all other countries:
Kluwer Academic Publishers Group
Distribution Centre
Post Office Box 322
3300 AH Dordrecht, THE NETHERLANDS

Library of Congress Cataloging-in-Publication Data

Heart function in health and disease : proceedings of the cardiovascular program sponsored by the Council of Cardiac Metabolism of the International Society and Federation of Cardiology during the Regional Meeting of the International Union of Physiological Sciences, Prague, Czechoslovakia, June 30 - July 5, 1991 / edited by Bohuslav Ostadal, Naranjan S. Dhalla.
p. cm. — (Developments in cardiovascular medicine ; DICM 140)
ISBN 0–7923–2052–2
1. Heart—Pathophysiology—Congresses. 2. Heart—Physiology--Congresses. I. Ostádal, Bohuslav. II. Dhalla, Naranjan S. III. Council on Cardiac Metabolism. IV. International Union of Physiological Sciences. Regional Meeting (1991 : Prague, Czechoslovakia) V. Series: Developments in cardiovascular medicine ; v. 140.
[DNLM: 1. Heart—physiology—congresses. W1 DE997VME v. 140]
RC682.9.H4 1993
616.1'2071—dc20
DNLM/DLC
for Library of Congress 92–48911
CIP

Printed in the United States of America.

This book is dedicated to

Professor Otakar Poupa

for his distinguished contribution to
Developmental and Comparative Cardiology.

Contents

Preface - A tribute to Otakar Poupa on his 75th birthday

Born in the hilly landscape of Eastern Bohemia on 17th October 1916, his early interests oscillated between biology and plastic art. Following the family tradition, he started his medical studies at the Charles University in Prague in 1935. His interest in basic sciences led him to work in the Department of Physiology headed by Prof. V. Laufberger, who discovered the ferritine, iron containing protein. Still as medical student, Poupa published his first paper dealing with the modification of the vascular effects of catecholamines by sexual hormones.

When the Charles University in Prague was closed during the German occupation (17th November, 1939) and research work was prohibited, he continued his work in endocrinology in a laboratory supported by the pharmaceutical industry. After Second World War, he returned to the Department of Physiology (in 1945). He started experiments on the metabolic antagonism of structural analogues (antithyroid drugs) which led him to study their interactions with nucleic acids (pyrimidine analogues). He performed similar investigations on histamine and its structural analogues. The results were published in two monographs and formed the basis of his postdoctoral thesis (1947). Soon thereafter he became Associate Professor of Physiology.

Unfortunately, after communist coup d'etat in 1948, Dr. Poupa was expelled from his university position and permitted only limited research facilities. During the political "thaw period", he was allowed to resume his academic position (1959) and, after 1961, became full Professor of Pathological Physiology at the Faculty of Pediatrics of the Charles University in Prague. There and in his laboratory at the Institute of Physiology, Czechoslovak Academy of Sciences, he started his studies on the onto- and phylogenesis of the cardiac muscle for understanding the basis of fetal and neonatal cardiology. This area of cardiological research, attracted many young students and became a speciality of the "Prague School" of developmental and comparative cardiology. These fruitful years

Dr. Otakar Poupa

(1959-1968) culminated during the "Prague Spring" 1968, when Dr. Poupa was elected as Corresponding Member of the Czechoslovak Academy of Sciences and honoured by a State Prize for his basic studies in cardiology.

For a short period in his life, he was active in politics. As one of the four authors of the heretic "2000 Words Manifesto" (together with macromolecular chemist O. Wichterle, cardiologist J. Brod and writer L. Vaculik), he emigrated after the Soviet occupation of Czechoslovakia in August 1968 and settled in Scandinavia for two decades. First in Sweden (Goteborg - A. Carlsten, B. Folkow) as Fellow of the Swedish Medical Research Council, in Denmark (Aarhus - K. Johansson) and in Norway (Bergen - K. Helle) as Visiting Professor. He found there favourable conditions for comparative cardiological studies and was able to continue his research on the hearts of rare and "unorthodox" vertebrates from which he tried to reconstruct the natural history of the heart ("The Heart Story", summary of this work, is published in the present volume). In 1976, on his 60th birthday, he was honoured for his research in comparative cardiology by an Honorary Doctorate of the University of Goteborg.

Otakar Poupa was also closely associated with the foundation on International Study Group for Research of Cardiac Metabolism (later International Society for Heart Research), the project which was initiated in Prague in 1964 during the International Congress of Cardiology (together with R. Bing and E. Bajusz). Otakar Poupa obtained the honorary membership of this Society in 1976.

The "velvet revolution" in November 1989 allowed him to visit Czechoslovakia again. On the occasion of his 75th birthday he was awarded the highest scientific assessment of his native country "The Merit of Sciences and Humanity" J.E. Purkyne and the "Golden Medal of the Masaryk University" in Brno. Despite his difficult life and scientific history, Dr. Poupa published 260 scientific papers, 6 monographs and many essays dealing with cultural life. The best evidence of his scientific personality are, however, the many pupils-experimental cardiologists working not only in Czechoslovakia, but also in many European countries, as well as on the American continent.

We wish him many more fruitful years in his beloved field, to which he has dedicated most of his life's work. And, of course, in his favourite hobby, painting, in which he has attained creditable renown. This book, based on some selected articles from the Cardiovascular Program at the Regional Meeting of the International Union of Physiological Sciences, has been put together in honor of Dr. O. Poupa to celebrate his achievements in heart research and recognize his stimulating leadership in the field of experimental cardiology.

It has been the lifetime desire of Dr. Poupa to clearly define the scientific basis of cardiology. He focussed his attention in this area because early on he recognized that heart disease is a major killer in the Western world and its trend is on the rise in developing countries. As the majority of cardiovascular deaths are related to myocardial ischemia, it is considered crucial to understand various aspects of ischemic heart disease. In this regard, it is noteworthy that ischemic heart disease is commonly associated with atherosclerosis, coronary spasm, as well as thrombosis leading to the development of arrhythmias, cardiovascular cell damage, myocardial infarction, cardiac hypertrophy and congestive heart failure. Furthermore, it is also important to appreciate various physiological, electrophysiological and biochemical processes in the normal heart if we are to understand their significance under pathological situations. This book containing 24 chapters has been organized in five sections to provide an outline of a complex problem in a convenient manner. One section of this book is devoted to shedding light on the restructuring and ontogenetic changes in the developing heart whereas in the next section some hypertrophic alterations due to chronic hypoxia are described. The third and fourth sections of this book are concerned with the regulation of cardiac channels as well as signal transduction mechanisms and cardiac electric field and these impart a comprehensive knowledge in the field of arrhythmias. The fifth section contains some pathophysiological events during the development of cardiac hypertrophy and heart failure. All these areas encompass a significant body of new information that should be valuable to those who work in the field of cardiovascular sciences, as well as those who are engaged in treating patients with heart disease.

Bohuslav Ostadal
Prague, Czechoslovakia

Naranjan S. Dhalla
Winnipeg, Canada

Acknowledgements

Our sincere thanks are due to the Members of the Organizing Committee for the Regional Meeting of the International Union of Physiological Sciences, Prague and Dr. Stanislav Tucek, Chairman of the Programme Committee for their help in promoting cardiovascular activity at this meeting. The cooperation of Dr. Robert Jennings, Chairman of the Council of Cardiac Metabolism of the International Society and Federation of Cardiology, helped us to mount a high quality cardiovascular scientific program by well established investigators. The efforts of Drs. I. Ostadalova, F. Kolar, V. Pelouch and other members of the Institute of Physiology, Czechoslovak Academy of Sciences during this conference contributed greatly to the success of this project. The help of Florence Willerton and Mary Brown for the preparation of this book is highly appreciated. Special thanks are also due to the editorial staff of Kluwer Academic Publishers, Boston, for the care and interest in compiling this book.

A. THE DEVELOPING HEART

Heart Story: A View To the Past

O. Poupa

Professor Emeritus, Göteborg, SWEDEN

Planet Earth is inhabited by 1,070,174 identified animal species, but only 6-7% of them are Chordates with closed circulatory circuit and cardiac pump. Chordates appeared 500 million years ago in paleozoic ocean and are called Agnatha.

Low-and High-Effective Pumps

Hagfish (Myxine glutinosa) one of such surviving fossils is equipped by two hearts, which pump almost completely desaturated blood with extremely low O_2-capacity (1 ml O_2.100 ml^{-1}) and substantial P_{50} switch to the left (P_{50} = 5) which implies that this cyclostome operates under severe anaerobic conditions. Its low performance pump (MV = 1 ml.min^{-1}; 15-20 beat.min^{-1}) is equipped by mitochondria with poor network of cristae but strong glycolytic capacity which provides an enormous resistance to anoxia. Glycerides are preferential fuel and fatty acids (FA) are refused (1). In contrast, the most effective pumps occur in very small mammals (shrews) and birds (hummingbirds) of body weight around 2 g. Hummingbird Beija flor has to fly when sucking nectar from flowers as his essential food, consuming 42 ml O_2.g^{-1}.h^{-1}. Its blood capacity is 22 vol % of O_2 which means that MV = 12.6 ml.min^{-1}, when O_2 utilization is set to 60% (2). Hummingbird's heart has to transport each minute the weight of blood which is twelve times greater than the weight of its body by more than 1000 beats.min^{-1}. Fat, its preferred fuel is located as droplets in vicinity of mitochondria. O_2 and fuel dependence is so high that when continuous refuelling at the sunset is stopped by closure of flower crowns, hummingbird-the "overheated engine" stops to work, O_2 consumption drops to 0.39 $ml.g^{-1}.h^{-1}$ and the splendid animal falls in torpor seen in lizards and other poikilotherms. These extremes are reflected in subcellular construction of cardiocytes. Volume occupied by respiratory energy-releasing system (mitochondria-MIT) varies from less than 15 to 80 vol % and is closely related with heart beat frequency (Fig. 1). MIT-volume seems to be species-constant but can be increased by chronic hypoxia (simulated high altitude - 49) within 30% limit due to MIT own DNA which are, however, under higher command of nuclear DNA.

From Lacunae to the Coronaries

Relative heart weight (RHW) i.e. proportion of the body weight (BW) occupied by the heart weight (HW) differs in vertebrates with 1:20 ratio. The smallest is in bottom-bound fish (plaice) and highest in hovering birds. When HW is plotted against BW one obtains four basic "channels" (poikilotherms, small and big mammals and birds - 3) which contain also animals of different classes: tuna HW is located in small mammalian channel (4) together with large snakes (5) while large mammals (4000 g) belong to the bird channel. All channels have slopes below one which means that small animals have relatively larger hearts than big animals. Physically active vertebrates -poikilo and homoiotherms- have greater RHW than inactive ones (6) and RHW is also higher in wild than domesticated forms, which appears after weaning (7). Heart growth, according to experiments with labelled leucine is stimulated preferentially by resistance work (8) and by physical activity (7) especially against gravitation (9).

First coronary vessels appeared in some fish 395 million years ago. Their appearance is bound to transformation of cardiac musculature from spongy to compact texture (3,5,7,9). Spongy musculature is supplied from voluminous lacunae by a large amount of desaturated venous blood. Coronary arteries contain oxygenated blood which in fish represents only one hundreth of cardiac output (10) and is in contrast to the established coronary circuit in mammals where

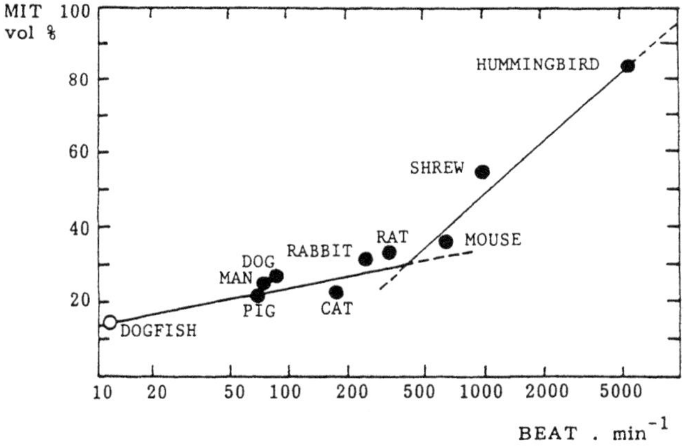

Fig. 1. Mitochondria in cardiac cell. Abscissa: cardiac beat frequency (min $^{-1}$). Ordinate: volume % of cell occupied by mitochondria.

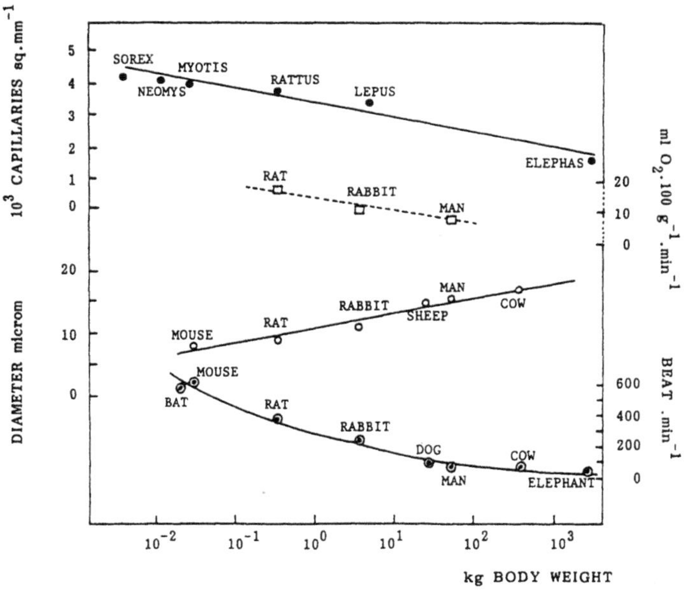

Fig. 2. Relationship between body weight (abscissa) and number of capillaries (10^3.sq mm $^{-1}$), O_2 consumption (ml.100g $^{-1}$.min $^{-1}$) of cardiac tissue, diameter of cardiac fibres (microm) and beat frequency (min^{-1}) of heart (5).

it is 4-5%. Among fish the proportion of cardiac wall occupied by coronary, supplied compact myocardium, varies considerably between 7-73% (11) and the highest figures occur in very active species (tuna, scomber, salmon, spratt, herring etc.). In developing salmon the compact compartment follows the volume increase of the ventricle except during smoltification (i.e. metamorphosis of the fish in sweet water preparing its trip to the ocean) when the growth of compact compartment and coronary capillaries disproportionately accelerates (12). In reptiles, there is no extra change in the compact layer size as a function of BW 0.34 g to 19.5 kg (5). Complete transformation of cardiac wall musculature with definitive installation of coronary vascular bed is achieved in mammals which took in the history of heart 170 million years. What triggers this transformation and mechanisms of the formation of coronary vascular bed are unknown. Studies on analogous transformation of chicken embryonal heart have shown the rise of mitotic endothelial activity in the wall of lacunae when they are compressed by growing muscular tissue followed by formation of first capillaries supplying condensing myocardial wall (13,14). Density of capillaries in mammalian myocardium decreases with body weight - in field-vole is twice that in elephant - and follows the trend of O_2-consumption of cardiac muscle (Fig. 2) (5). The size of cardiac fibres - in elephant twice that of the field-vole - is directly proportional to the stroke-volume and inversely proportional to the beat frequency, whereas the density of capillaries is inversely proportional to the size of the fibres but directly proportional to the beat frequency of the heart (5).

Cardiac muscle is supplied by nutrients and O_2 when relaxed and devoid of it at work. It follows that the stay of oxygenated hemoglobin is time-limited by beat frequency as well as length of diastole which varies with the BW. This makes O_2-unloading ability of blood (P_{50}) of considerable importance. These relationships are shown in Fig. 3, where the length of the diastole (beat frequency) is plotted against P_{50} of respective animal. In general, with shortening of the diastole P_{50} shifts to the right. At the left side of the plot are poikilotherms with low-frequency heart supplied

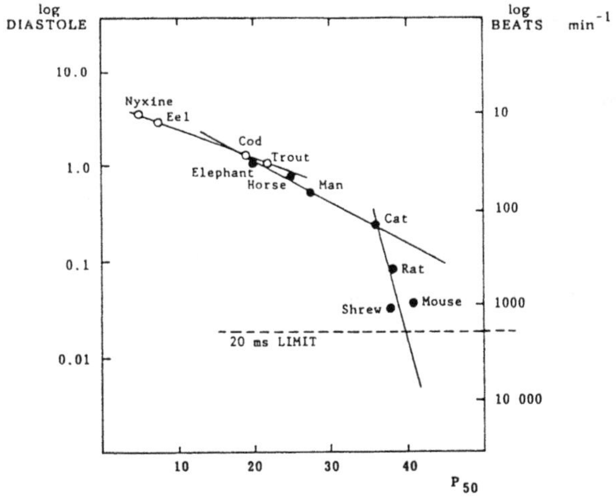

Fig. 3. Relationship between length of diastole (log s^{-1} resp. beat frequency in min^{-1} -ordinate) and blood P_{50} in different vertebrates (poikilotherms: open circles, homoiotherms: black points). 20 ms = velocity of Hb-HbO_2 dissociation.

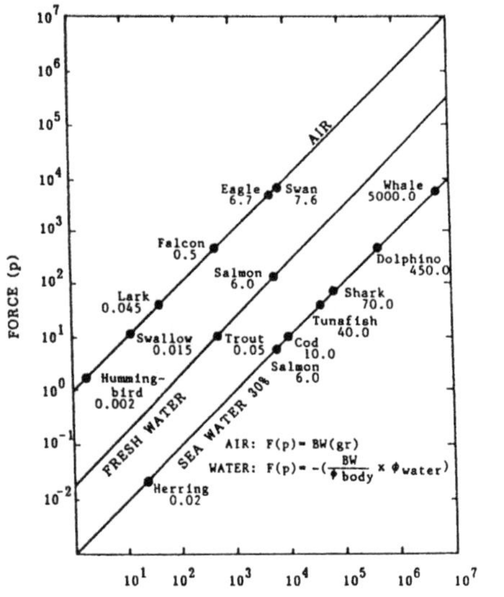

Fig. 4. Body weight (log g - abscissa) and antigravity force (log p - ordinate) relations. Plots demonstrate the force necessary to keep a body of certain weight (g) floating in different media (air, fresh and salt water). Body weight of some vertebrates (in kg) is noted (9,5).

by venous blood from lacunae, where Hb unloads at lower PO_2. Analogous situation occurs during long diastole in slow-beating heart in great mammals when unloading occurs at low PO_2. With increasing frequency O_2 is delivered at higher PO_2 (P_{50}) as the capillary blood contact with the tissue is shorter due to shorter diastole. Above 300-400 beat.min^{-1} the blood P_{50} does not correlate with the length of the diastole. It is of interest that within this range of heart beat the MIT volume sharply increases (Fig. 1) which indicates that the O_2-gradient is enhanced mainly by rapid O_2 consumption (16). As the reaction time of HbO_2 dissociation is 15-20 ms; this critical time interval, i.e. length of diastole occurs around 2000 beat.min^{-1}.

Water to Land Transition

A memorable step in the history of terrestrial vertebrates was their shift from the ocean to the land. It was preceded by land invasion of plants which, thanks to their chloroplasts, enriched the atmosphere by oxygen and made the land inhabitable for animals. This all occurred 350-400 million years ago. O_2 content in the atmosphere - although only 50% of today - was high when compared with the sea water. This allowed to reduce substantially ventilation, which on the other hand, exposed them to CO_2 retention and acidosis as undesirable consequence. Blood pH dropped by 0.5-0.7 units in spite of increased buffering. A further new factor was the impact of gravitation which made transport of the body more expensive so that first land-dwellers had often to work on the O_2-debt paid sometimes with considerable delay. A third new factor was a large spectrum and variations of the air temperature when compared with the water environment. To escape to the temperature comfortable zone - as seen in water-dwellers - was difficult or impossible. The solution was either to adapt metabolic functions to environmental temperature or to create its own thermostatic systems. In all these three new features was - directly or indirectly - cardiac muscle involved.

Gravitation

The demand which the antigravity effort exerts on the muscular

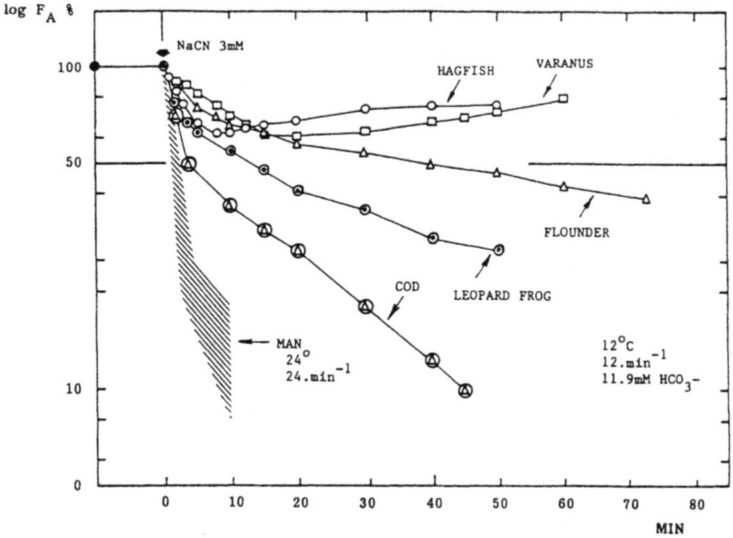

Fig. 5. Cardiac force under anaerobiosis. Force decline (in % of F - ordinate) of cardiac strips in vitro (12°C paced 12 min in Ringer with 11.9 mM HCO_3-; man: 24 °C paced 4 min^{-1}) when respiratory energy production was stopped by CN^- (3 mM). Abscissa: min.

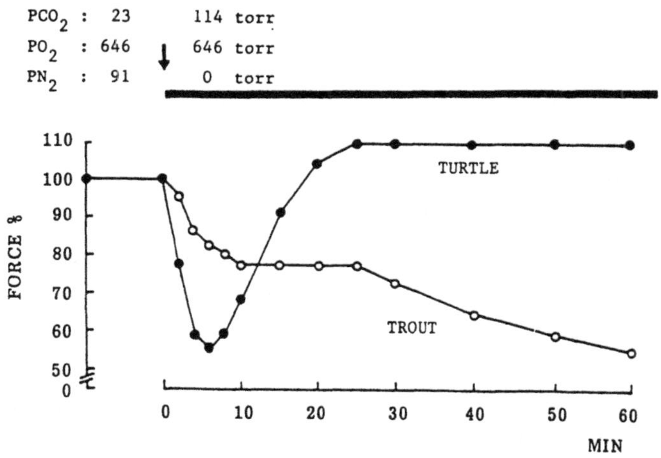

Fig. 6. Effect of respiratory acidosis on cardiac force in water (trout) and terrestrial vertebrate (turtle). Ordinate: force decline of cardiac strip (in % of F_A) PCO_2 - change from 23 to 114 torr at arrow. PO_2 constant (646 torr). Abscissa: time in min. (17).

activity is illustrated by the plot in Fig. 4, which shows the relation between the BW and the force necessary to keep the body floating in different media (air, fresh and salt water). Keeping a motionless body of five tons within sea water (e.g. whale) entails development of force of the same order required to keep a body of 6 to 10 kg (eagle) hovering in the air. For birds the raising of the center of gravity requires a power proportional to the length raised to the power of 3.5, whereas for fish this work can be virtually neglected (5). This presumption reflects in size of cardiac muscle: under transition from water to land the heart has more than tripled its size (fish RHW: 0.081 ± 0.03 and amphibian RHW: 0.299 ± 0.17). When homoiotherm had to fly the RHW was almost doubled (mammals: 0.643 ± 0.19, birds: 1,099 ± 0.32 - 3,5,7,9).

Retention of Carbon-Dioxide

Due to the high content of oxygen in the air, first inhabitants of the land were able to decrease ventilatory effort and to use the lungs as new gas exchanger. The problem of how to get oxygen was solved, but a new problem appeared: how to cope with CO_2 and H^+ excess. The situation was even worse in reptiles because they could not eliminate CO_2 through the skin as amphibia due to impermeable shells and scales. In reptilians great part of the cardiac wall is still spongy supplied from lacunae by mixed or venous blood not seldom containing high concentration of lactic acid due to slowly paid O_2-debt. All of this is the probable cause of why reptilian heart is champion and how to cope with anoxia and acidosis. Plots in Fig. 5 compare force development of cardiac muscle strips from different vertebrates when MIT-activity is blocked. Varanus displays enormous resistance similar to other members of reptilian class. Fig. 6 shows difference of inotropic reaction to increased CO_2 between fish and reptilian heart: in trout peak tension steadily decreases, meanwhile turtle heart strip after initial depression, regains force in spite of ongoing acidosis (17). This ability of the heart to work at high PCO_2 has developed parallel with fluid and tissue CO_2-content (18,19,20) and cannot be explained either by different tissue pH (21) or buffering capacity (22). As Ca^{2+} is a

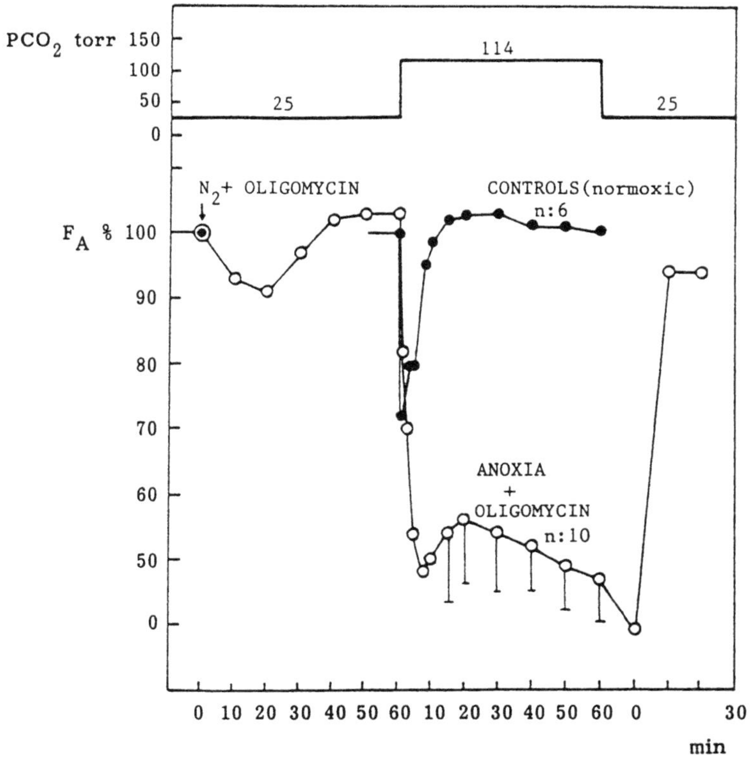

Fig. 7. Prevention of spontaneous force recovery under ongoing acidosis in terrestrial vertebrate heart. Isometric cardiac strips of viper. Black points: untreated controls (normoxic). Open circles: strips pretreated by N_2 + oligomycine for 60 min. Upper plot: PCO_2 in torr. Ordinate: force in % of F_A. Abscissa: time in min. (20).

known H^+ competitor at active spots of cardiac cell, one can propose that CO_2-depressant effect is counteracted by Ca^{2+} activation. Because enhanced Ca^{2+} sarcolemmal influx or SR as Ca^{2+} source were excluded (20), mitochondrial Ca^{2+} was proposed as tentative candidate: by reversible block of mitochondrial Ca^{2+}-Na^+ exchange (anoxia and oligomycine) in reptilian heart the force recovery under ongoing CO_2 acidosis was prevented (see Fig. 7; 20).

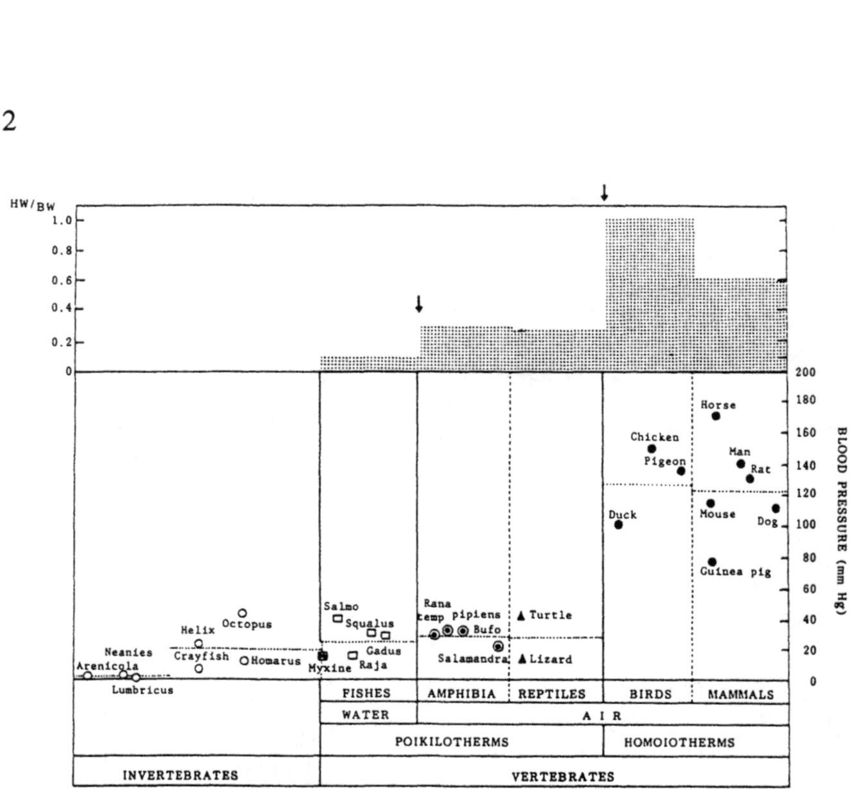

Fig. 8. Relative heart weight (HW/BW) and blood pressure (mm Hg) during evolution of vertebrates.

Thermoregulation

Water offers to the inhabitants more stable thermal conditions with lesser peak values (water 0-30 °C; air -70 to +50 °C). When necessary it allows less expensive transport by swimming to more comfortable temperature zones. Also the access to nutrition is easier at least for smaller fish with modest food intake. Big, voracious, carnivorous fish (sharks, tunas etc.) during pursuit of their prey are using more effective red musculature around the spine, which consumes six times more O_2 than white muscle segments. This creates steep thermal gradient between core (30-32°C) and surface of the body (20-22°C), which is kept by counter-current vascular rate, where cooler arterial blood from grills is preheated by warmer venous blood returning from deep red muscles. This elaborate system of vessels creates extra peripheral resistance and is probably the cause of high RHW in tunas which transfers them from fish channel to small

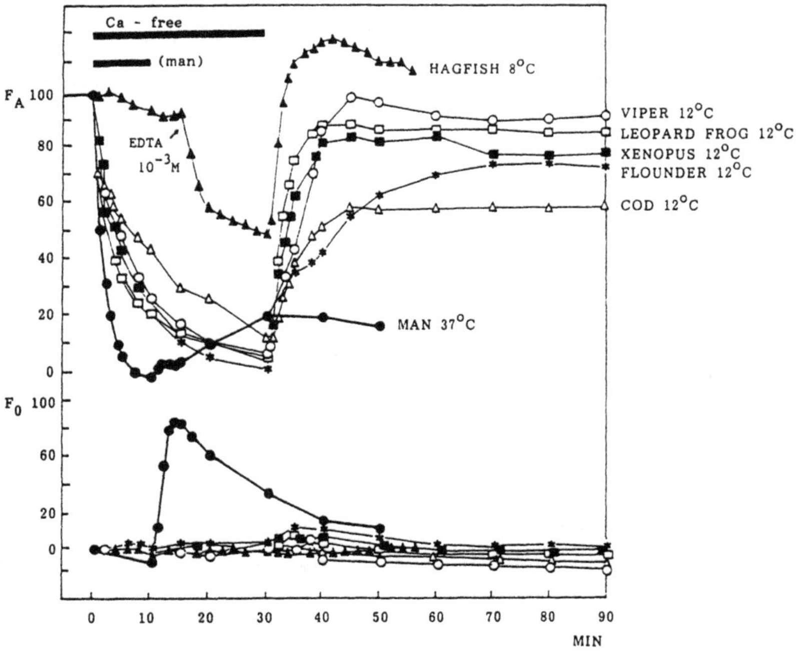

Fig. 9. "Calcium paradox" in different vertebrates. Force recovery (in % of initial F_A - upper plot) and initial tension (F_0 - lower plot) after exposure of isometric cardiac strip to Ca^{2+} - free Ringer and subsequent Ca^{2+} - Ringer reinfusion (30).

mammalian channel when HW is plotted against BW (4). Analogous feature occurs when comparing RHW in homoio- and poikilotherms. Thermoregulating mammals and birds have in the mean RHW four times and blood pressure six times higher than poikilotherms (Fig. 8). Nevertheless, they are also other causative factors as higher blood viscosity, fully developed pulmonary circuit etc. However, shifts of blood volume between body surface and core are the main thermoregulatory mechanism keeping body temperature stable within thermoneutral climatic zone. When crossing its lower critical point - which is the rule in temperate climate - the perfusion of body surface must be limited by the flow-resistance to prevent the loss of heat. This explains why identical homoiotherms living in the northern part of the continents have higher RHW than in warmer southern climate (24). Insulation - important factor preventing the heat loss - becomes the risk-factor when the temperature crosses the high critical point. Deleterious could be sudden discharge of catecholamines probably due to their calorigenic effect which increases metabolic rate up to 30%. It was shown that in animals equipped by fur (rats) the cardiotoxicity of beta-agonists substantially increases by elevated environmental temperature which could be prevented by clipping the fur (25). Restrained stress releasing endogenous catecholamines in rats which have no subcutaneous fat layer produces cardiac lesions of only minor importance (26), whereas in animals with continuous fat-layer (pigs) even a short restraint produces severe large ventricular necroses not seldom with a fatal outcome (27). Vulnerability of cardiac cell largely depends on function of sarcolemma and its integrity. This is shown by "Ca-paradox". When homoiotherm cardiac muscle is exposed to temporary Ca^{2+}-lack and reperfused by Ca^{2+} again, sarcolemmal phospholipid bilayer is damaged and cardiac cell dies displaying micro changes similar to cardiac necrosis (28). This pathology cannot be produced in poikilotherms (29,30). In hagfish the absence of Ca paradox is a special case. Cardiac cell displays a remarkable lack of sensitivity to extracellular Ca^{2+}, probably due to large Ca-store within voluminous glycocalyx (30), which operates as a Ca-buffer zone. As Ca-paradox can be protected in homoiotherms by cooling (31,32,33) it was suggested that the lack of it in poikilotherms is due to the lower temperature. However, this proposition could not be proven (29): even at substantially elevated temperature the Ca-paradox cannot be provoked. When cellular damage is quantified as the extent of force recovery after Ca^{2+}-reinfusion and expressed by Arrhenius plot, a sudden change in activation energy occurs at 23°C (Fig. 9). Analogous change in activation energy was found in the activity of membrane-bound enzymes which correlates with temperature-induced phase changes in the lipid components of membranes (34). Poikilotherm membranes contain more polyunsaturated FA than homoiotherm, which indicates protective effect of

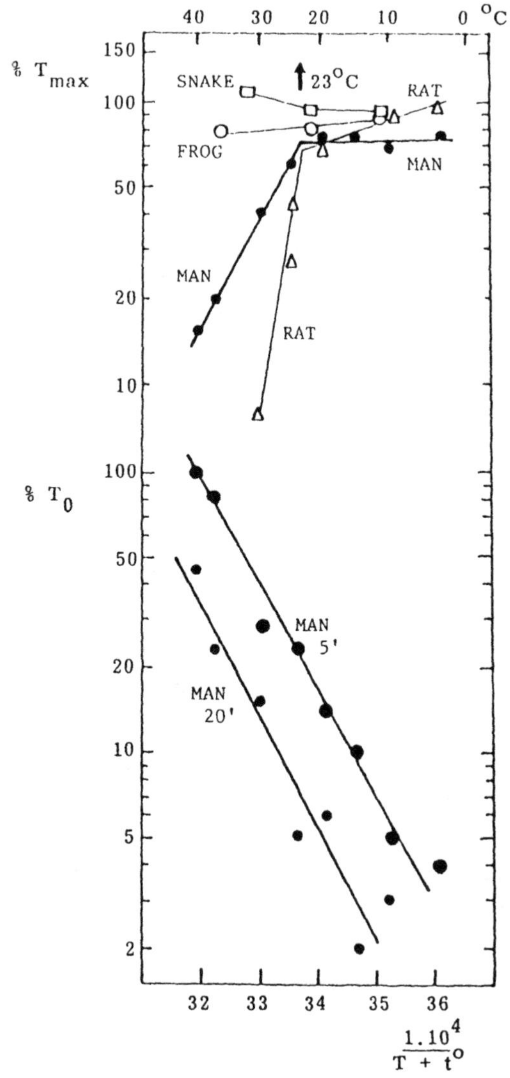

Fig. 10. Temperature and "calcium paradox" in homoiotherms (rat, man) and poikilotherms (viper, frog) expressed by Arrhenius plot. Upper plot: F_A, Lower plot: F_0.

polyunsaturated FA on cardiac sarcolemma, which makes its integrity less Ca^{2+} -dependent. This hypothesis deserves experimental verification.

Notes to Individual Development

In early stage of cardiogenesis, the heart is a tubular rudiment containing three components (35): outer myocardium with early contracting myofibrils, endocardium and between them cardiac jelly, which develops in fibrous matrix containing terminal capillary network and coronary vessels in homoiotherms. Homoiotherm heart muscle fibres cease dividing early in postnatal life (36), whereas stromatic cells retain their ability to divide through the life span. Also capillarogenesis proceeds later than myofibrillogenesis. This is seen in fibre to capillary ratio, which at birth in mammals is four to one, but in adulthood one to one found in nine different mammalian species (37,38,39,5). It follows that under overload the cardiac growth is achieved by increased volume of myofibres (hypertrophy) (3,7) with increased diffusion distance as the consequence. Cardiac hyperplasia was shown when overload occurred early in postnatal life and was caused by volume overload (anemia - 41). In wild animals the increased RHW displays higher density of capillaries and smaller size of myofibres when compared with domesticated partners (37,38). Also the stromatic component is lower in wild rats compared with laboratory rats (38). Most prominent is higher capillary density and smaller size of myofibres in flying mammal (bat) when compared with BW-matched animal (mice) but lower RHW (39). Nothing is known about cardiac growth mechanisms in poikilotherms but some indirect findings indicate that the growth capacity of myocardium and its stroma differs from homoiotherms. In homoiotherms, the injury of muscle fibres is followed by activation of stromatic components, i.e. fibroblasts and capillaries and to the end the defect is replaced by fibrous scar. Quite different is the situation in some poikilotherms. In amphibia (frogs) the stromatic activation and infiltration by leucocytes after severe injury is missing (42,45) and large primary aneurysms produced either by catecholamines or elevated temperature are never replaced by fibrous

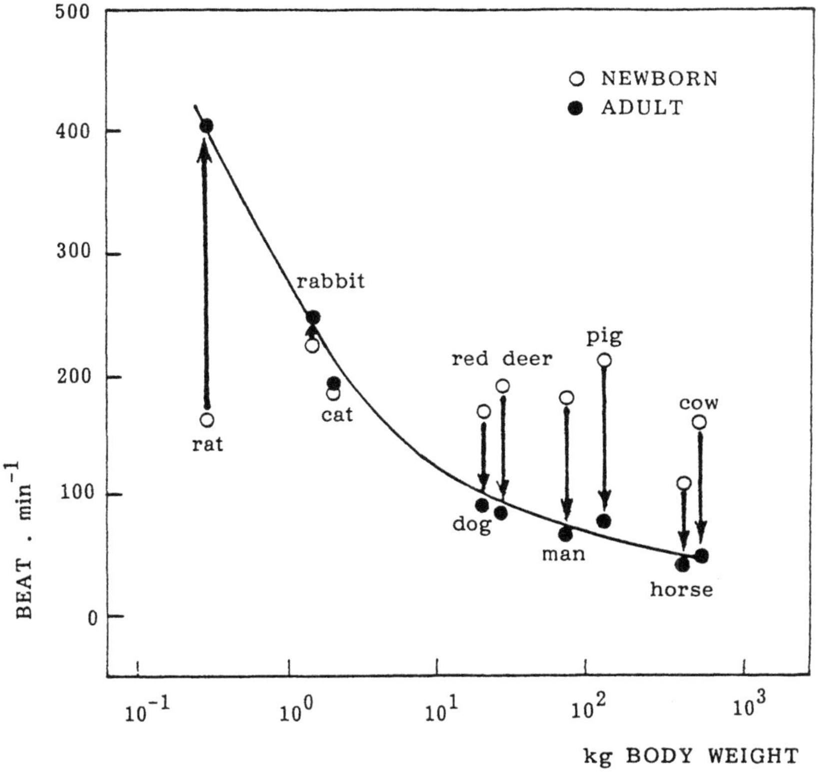

Fig. 11. Cardiac beat frequency (min^{-1}-ordinate) in newborn (open circles) and adult (black points) mammals of different body weight (log kg - abscissa).

scars (42,44,45,46,42). In reptiles (turtle) catecholamine-induced cardiac lesions are similar as in homoiotherms: coagulation necrosis is followed by infiltration and activation of stromatic elements (47,44).

Transition from fetal to newborn life is one of the most dramatic experiences under individual development. Because delivery occurs in homoiotherms at different stage of cardiogenesis, caution should be taken when extrapolating from one to another species. Delivery used to be compared with descent from high altitude to the sea level. This verified measurements of cardiac muscle anaerobic

Table 1.

	Beat min^{-1}	Length of cycle	Length of diastole	MV	SV
MOUSE	615	98 ms	59 ms	4.1 ml	0.007 ml
WHALE	15	4 s	2.4 s	1237 l	82 l

Table 2.

SEMANTIC OF BODY ORGANS

(The concise Oxford Dictionary)
7th Edition 1983

ENTRY	NUMBER OF LINES
Eye	90
Heart	81
Blood	61
Skin	42
Ear	39
Breast	33
Stomach	30
Brain	29
Kidney	13
Lung	13
Liver	12
Muscle	11

capacity in patients under reconstruction of Fallot's tetralogy with PO_2 are similar to intrauterine values or in high altitude dwellers (personal communication of M. Kopetzky and S. Daum), which displayed similar values obtained earlier in high altitude adapted animals (49,50,51,9). In the growing organism, cardiac performance has to adapt to demands of O_2 and nutrients of body mass which increases in different species by various velocity, reaching terminal plateau from grams to tons with various weight-specific metabolic rates which differ between mice and elephant as 24:1 (mouse 1.65 and elephant 0.07 $O_2 .ml.g^{-1} .h^{-1}$). Table 1 gives basic parameters of cardiac activity in adult mouse and adult whale. It shows that small mammal solves the transport of blood by low volume and high speed pump, whereas big animal has high volume and low speed pump. This is in contrast to fetal life where cardiac beat frequency does not correlate with body size of respective species being around 150-200 $beat.min^{-1}$. The plot in Fig. 11 shows that in postnatal development, mammals with BW under 2000-3000 g accelerate cardiac frequency, whereas in heavier mammals, heart beat decelerates. This indicates preponderance of vagal tonus in big and sympathetic drive in small mammals. As men belong to heavy mammals, their preponderant cardiac innervation seems to be vagal with all risks of enhanced sympaticus drive to which expose him daily to stressors of a contemporary affluent, overheated, industrial society.

Epilogue

100,000 years ago on our planet appeared a man - Homo Sapiens - and it took him ninety eight thousand, three hundred and seventy two years until in England, William Harvey discovered that the heart is a pump. But this honorable medicus of kings hesitated to publish his discovery: since centuries heart became a sacred organ and only heretic dares to compare it with profane pump. When looking at Table 2 which tries to express the semantic of body organs by the count of lines in Oxford Dictionary, one can see the heart at the top of the list. For a romantic poet, it is the seat of passions both marvellous and painful and for the man of the street it is the watch indicating by its beat his time on the earth. Let us allow a futurological question: in the history of life the heart appeared and as any historical event, it will also disappear. When will the heart disappear as an unnecessary organ? Neurophysiologists of today try to construct artificial intelligence. When they will succeed to replace human brain by system of microchips the heart will become unnecessary because microchips need neither oxygen nor fuel: they are deadless. But at the same time, the birdsong and the gunfire will be nothing more than signals of different frequencies.

References

1. Sidell DB. Comp Biochem Physiol 1983;76A:495-505.
2. Johansen K. The world as a laboratory. A Krogh Memorial Lecture, 1986.
3. Poupa O, Ostadal B. Ann NY Acad Sci 1969;156:445-468.
4. Poupa O, Lindström L, Maresca A, Tota B. Comp Biochem Physiol 1981;70A:217-222.
5. Poupa O, Lindström L. Comp Biochem Physiol 1983;76A:413-421.
6. Clark AG. Comparative Physiology of the Heart, Cambridge Univ. Press 1927.
7. Poupa, O, Rakusan K, Ostadal B. In: Medicine and Sport (eds.) D Brunner, E Jokl. Kargel Basel 1970;4:202-233.
8. Hjalmarson A, Isaksson O. Acta Physiol Scand 1972;86:126-144.
9. Poupa O. In: Myocardiology. Recent Advances in Studies of Cardiac Structure and Metabolism (eds.) G Rona, E Bajusz. Univ. Park Press 1972;1:779-802.
10. Belaud A, Peyraud C. J Physiol 1971;63:165A.
11. Santer RM, Greer-Walker M, Emerson L, Witthames PR. Comp Biochem Physiol 1983;76A:453-457.
12. Poupa O, Gesser H, Johnson S, Sullivan L. Comp Biochem Physiol 1974;48A:85-95.
13. Rychterova V. Folia Morphologica 1971;19:262-272.
14. Rychterova V. Folia Morphologica 1977;25:7-14.
15. Loiselle DS, Gibbs CL. Amer J Physiol: Heart Circ Physiol: 1979;6(1):H90-H98.
16. Poupa O, Brix O. Comp Biochem Physiol 1984;76A:1-3.

17. Poupa O, Gesser H, Johansen K. Amer J Physiol 1977;234:R115-R157.
18. Poupa O, Johansen K. Amer J Physiol 1975;228:684-688.
19. Gesser H, Poupa O. J Comp Physiol 1978;127:307-313.
20. Gesser H, Poupa O. Comp Biochem Physiol 1983;76A:559-566.
21. Gesser H, Jorgensen E. J Exp Biol 1982;96:405-412.
22. Damm Hansen H, Gesser H. J Exp Biol 1980;84:161-167.
23. Cloud P. Scientific American 1983;249:132-149.
24. Hesse R. Zool J 1921;38:243-364.
25. Faltova E, Poupa O. Canad J Physiol Pharmacol 1969;47:295-299.
26. Selye H. The pluricausal cardiopathies. Ch. C Thomas Springfield 1961.
27. Johansson G, Jönsson L, Lannek N, Blomgren L, Lindberg P, Poupa O. Amer Heart J 1974;87:451-457.
28. Zimmerman ANE, Hüsmann WC. Nature 1966;211:646-647.
29. Lagerstrand G, Mattisson A, Poupa O. Comp Biochem Physiol 1983;76A:601-613.
30. Poupa O, Helle K, Lomsky M. Comp Biochem Physiol 1985:81A:801-805.
31. Holland CE, Olson RE. J Mol Cell Cardiol 1975;7:912-928.
32. Hearse DJ, Humphrey SM, Bullock GR. J Mol Cell Cardiol 1978;10:641-668.
33. Lomsky M, Ekroth R, Poupa O. Europ Heart J 1983;4:(Suppl H)139-142.
34. McMurchie EJ, Raison JK, Cairncross KD. Comp Biochem Physiol 1973;44B:1017-1026.
35. Manasek FJ. In: Developmental and Physiological Correlates of Cardiac Muscle (eds.) M Lieberman, toyomi Sano. Raven Press 1976;1-20.
36. Fischman DA, Doyle CM, Zak R. In: Developmental and Physiological Correlates of Cardiac Muscle (eds.) M Lieberman, Toyomi Sano. Raven Press 1976;67-79.
37. Wachtlova M, Rakusan K, Poupa O. Physiol Bohemoslov 1965;14:328-331.
38. Wachtlova M, Rakusan K, Roth Z, Poupa O. Physiol Bohemoslov 1967;16:548-554.

39. Wachtlova M, Poupa O, Rakusan K. Physiol Bohemoslov 1970;19:491-495.

40. Roberts JP, Wearn JT. Amer Heart J 1941;21:617.

41. Poupa O, Korecky B, Krofta K, Rakusan K. Physiol Bohemoslov 1964;13:281-287.

42. Korb G, Poupa O, Carlsten A. J Mol Cell Cardiol 1973;5:313-317.

43. Poupa O, Carsten A. Amer Heart J 1970;80:843-844.

44. Poupa O, Carlsten A. In: Recent Advances in Studies of Cardiac Structure and Metabolism (eds.) E Bajusz, G Rona. Univ. Park Press 1973;2:321-351.

45. Carlsten A, Poupa O, Volkmann R. Comp Biochem Physiol 1983;76A:567-581.

46. Carlsten A, Ericson LE, Poupa O, Winell S. Comp Biochem Physiol 1983;76A:583-591.

47. Ostadal B, Rychterova V, Poupa O. Amer Heart J 1968;76:645-649.

48. Carlsten A, Ericson LE, Poupa O. Arch Pathol Microbiol Immunol Scand Sect A 1982;90:57-65.

49. Poupa O. In: Medizinische Aspekte der Höhe (eds.) P Deetjen, E Humpeler. Thieme Vrlg Stuttgart, N York, 1981;37-61.

50. Kopecky M, Daum S. Cs Fysiol 1958;7:518.

51. Poupa O, Krofta K, Prochazka J, Chvapil M. Physiol Bohemoslov 1965;14:233-236.

Postnatal Growth and Restructuring of Mammalian Hearts

K. Rakusan

Department of Physiology, Faculty of Medicine, University of Ottawa, Ottawa, Ontario, CANADA

Introduction

There are several good reasons for studying cardiac growth and development. The first impulse stems probably from natural curiosity, namely an attempt at "mapping of the new territory". Knowing the "developmental chart" is also useful in various fields of experimental and clinical cardiology for the following reason: the developmental stage determines cardiac response to various physiological and pathological stimuli, as documented in various contributions to this symposium. Finally, even cardiac failure may be considered an abnormality of cardiac growth. This is not a new idea. The notion that cardiac hypertrophy, as extended cardiac growth, contains seeds of cardiac failure, has already been recognized by Osler (1). Osler's analysis based on clinical observations was expanded subsequently by Meerson who used the experimental approach (2). Recently, a similar idea has been proposed by Katz (3,4). We will return to this topic at the end of this presentation.

Postnatal development of mammalian hearts will be illustrated by examples of changes at the organ level in terms of cardiac mass as well as at the tissue level, where the two important tissue components will be analyzed: cardiac myocytes and coronary blood capillaries. Most of the examples will originate either from studies on human or rat hearts.

Cardiac Mass

Postnatal development of cardiac mass in human hearts is summarized in Fig. 1. This is based on three sources. The oldest record of cardiac weight can be found in Natural History by Pliny the Elder, in the first century A.D. It is based on the experience of Egyptian embalmers who believed that the human heart grows larger every year, reaching its heaviest weight at the age of 50, and from that point on losing weight at the same rate. Consequently, cardiac weight is decreased to its minimal value by the age of 100 years, and the resulting atrophy of the senile heart was held responsible for the accepted life span of 100 years. Data in Fig. 1 for the ancient descriptions are based on a more detailed report of the same findings by Censorinus, a Roman astrologer, of the third century. Both descriptions can be found in "Origins of the Study of Human Growth" compiled by Boyd (5). These ancient measurements are compared with two contemporary reports: the study of Linzbach and Akuamoa-Boateng (6), and that of Olivetti and coworkers (7). The former study is based on the largest collection of data (112 human hearts) while the latter represents a smaller sample of carefully selected data which are from cases apparently free of diseases or symptoms affecting cardiac mass. Two interesting conclusions may be drawn from these comparisons:

1. Ancient descriptions for most of the life span are well within the accepted limits of the human cardiac weight as we know it in modern times (if one translates units of weight from drachma into grams).
2. According to Linzbach and Akuamoa-Boateng (6) the absolute cardiac weight increases continuously up to the ninth decade of life, with only a slight reduction during the 10th and 11th decade. On the other hand, in agreement with the ancients, Olivetti and coworkers (7) found a decrease in absolute cardiac weight with aging, albeit at a slower rate.

Postnatal development of cardiac weight is also characterized by a decreasing contribution of the right ventricle. This is also the case in various mammalian hearts we reviewed recently (8). In the same review, data for various experimental animals can be found.

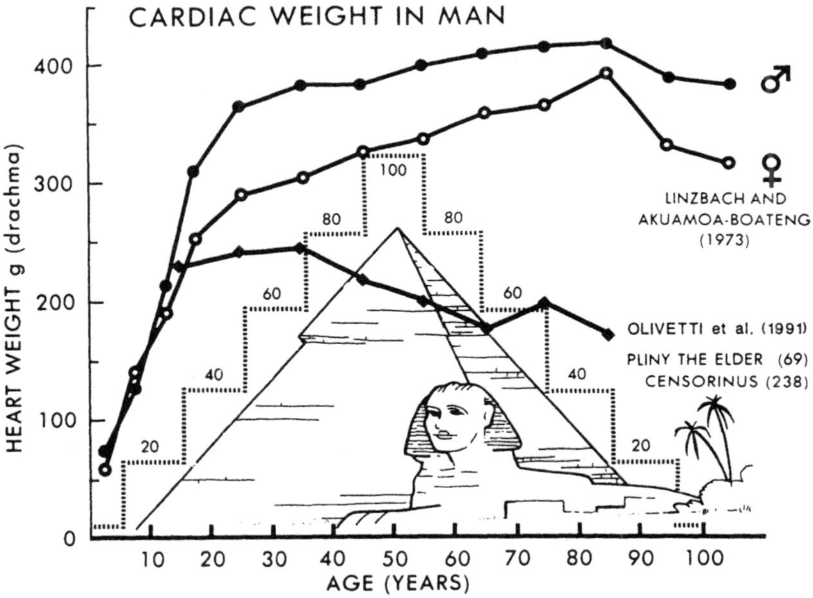

Fig. 1. Postnatal changes in cardiac weight in Man. Ancient data expressed in drachma and scaled to grams are taken from Boyd (5). They are compared with results of Linzbach and Akuamoa-Boateng (6) and Olivetti et al. (7).

Generally, they display a similar trend as in human hearts, characterized by decreasing relative heart weight with increasing body weight.

Cardiac Myocytes

Basically, an increase in cardiac mass may be due to hypertrophic or hyperplastic growth of myocytes or to a combination of both processes. The term hypertrophy indicates an increase in size of the individual myocytes without changing their total number, whereas in hyperplasia, proliferation of the myocytes occurs. Early postnatal growth of the mammalian heart is realized by an increasing volume of individual myocytes as well as an increasing number. The cell diameter, which appears to be similar in hearts from all mammalian species, increases during the life span to two to three

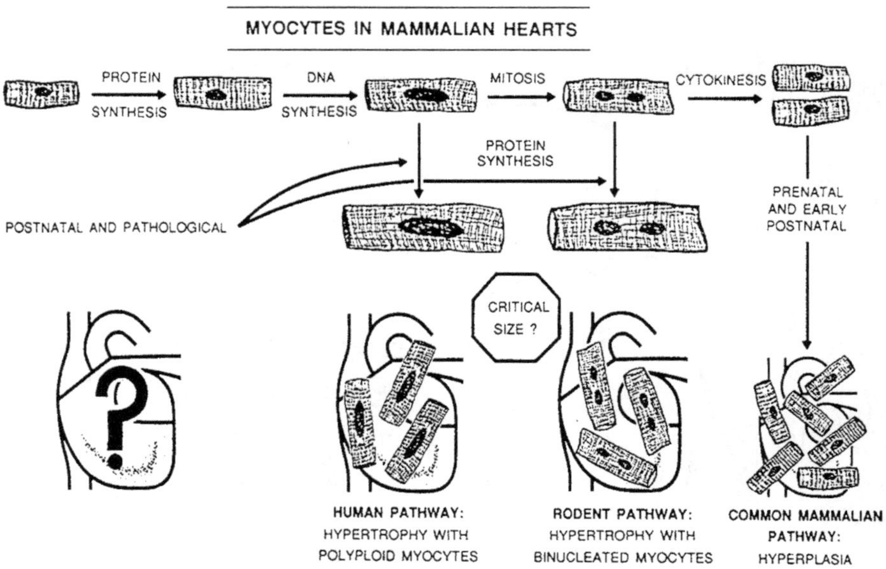

Fig. 2. Postnatal growth and development of cardiac myocytes. Based on results reviewed by Rakusan (8).

times the value at birth (8). In contrast to older concepts of the heart as a cell-constant organ, it seems evident that proliferation of muscle cells takes place in newborn hearts. It is not clear, however, when or why the proliferation stops. In the case of the rat heart for example, the "cut-off" point has been proposed as somewhere between the third and sixth postnatal weeks (8).

Regulation of the cardiomyocyte growth, differentiation and function is encoded by cellular oncogenes (c-oncs, protooncogenes). C-oncs are any genes which encode growth factors, receptors, intracellular signalling molecules or proteins that regulate transcription (9). There are at least 30 known growth factors - their review is beyond the scope of this presentation. It is important, however, to keep in mind that the ability of cardiac myocytes to proliferate is repressed rather than lost. A terminally differentiated phenotype is probably due more to the inhibitory actions of certain substances than to a loss of receptors for growth factors. In this case, repression or neutralization of the inhibitor should enable the cells to cycle again. The key role is probably played by the retinoblastoma gene product and cdc_2 kinase (10).

Possible modes of cardiac myocyte growth and development are schematically depicted in Fig. 2. The pre-natal and early postnatal periods can be characterized as follows: protein synthesis leads to an increase in mass, which is accompanied by DNA synthesis in nuclei and subsequently followed by nuclear and cell division (mitosis and cytokinesis). Thus, prenatal and early postnatal growth is realized mainly due to cell hyperplasia. This characteristic chain of development is probably common to all mammalian hearts. Later on, hyperplasia ceases to be a normal mode of growth and future growth is a result of an increase in the size of myocytes, hypertrophy. In this case, however, at least two distinct modes of actions may be distinguished. In the case of rats, the last stage is mitosis without accompanying cytokinesis, and the resulting myocytes are mainly binucleated or polynucleated. Subsequent growth is of the hypertrophic type, which is, however, limited by a critical size of the cells (11). Finally, myocytes from human hearts are predominantly polyploid, and similarly some critical size of the cells may also be assumed in this case. Figure 2 summarizing these three types of myocyte development is based on results reviewed by us recently (8).

Coronary Capillaries

Coronary capillaries are not fully developed at birth. Their maturation continues in the early postnatal period, it is characterized by establishment of definite morphological and biochemical features of the vascular walls as well as changes in the spatial arrangement of capillary orientation (12).

The early postnatal period is characterized by a rapid rate of coronary capillary formation. For instance, close to half of all adult capillaries in the rat heart are formed during the first 3-4 postnatal weeks (13). Similarly, Olivetti and coworkers (14) reported that during the first 10 postnatal days, capillaries grow

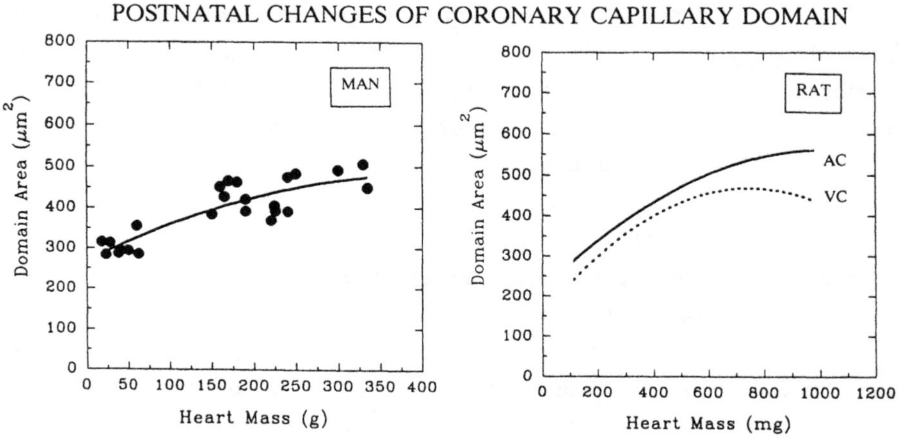

Fig. 3. Postnatal changes in coronary capillary domain. Data for human hearts originate from Rakusan et al. (16), for rat hearts from Batra and Rakusan (15). AC = arteriolar capillary domain, VC = venular capillary domain.

two to three times more rapidly than the myocardial mass.

The main quantitative features of the coronary capillary bed during the subsequent period are decreases in the fiber/capillary ratio and decreases in capillary density. For instance, the number of myocytes supplied by a single capillary decreases from 4-6 in the neonatal period to one in adult hearts from various mammalian species (man, rabbit, rat; 8). The decrease in capillary density with increasing age and cardiac mass is also present in all mammalian hearts. Fig. 3 summarizes postnatal development of the capillary domain in human and rat hearts. The capillary domain is an area supplied by a single capillary on a tissue cross-section. The area is delineated by a computer program which selects the region of tissue nearer to that capillary than any other. Figure 3 demonstrates several important features. First of all, domain area increases with increasing cardiac mass, reflecting decreasing capillary density. Secondly, the area is similar in both species of such different body size. It seems that not only is the myocyte size more or less the same in all mammalian hearts regardless of the body mass, but also the capillary supply area is approximately the same. Finally, domain area in the rat heart is subdivided using differential staining into the portions close to arterioles (AC) and those close to venules (VC). Arteriolar regions of capillaries with a presumably higher PO_2 supply a larger tissue area than do venular regions (15).

Conclusions

We have mentioned in the introduction that the knowledge of developmental stage will help to predict and explain the reaction of a heart to various physiological stimuli. To illustrate this thesis, let us consider pathological growth of the heart due to pressure and/or volume overload. Results will be different if the growth stimulus is applied in a young growing organism than in an adult. Previous studies on postnatal development of cardiac mass, coronary capillaries and cardiac myocytes will explain why, for instance, pressure overload introduced in the neonatal stage will result in a larger increase in cardiac mass (due to both hypertrophy and hyperplasia of cardiac myocytes and commensurate capillary growth) when compared to the results of the same stimulus in the adult organism. This is true for human hearts as well as for rat hearts (16-18).

Transition from pathological cardiac growth to cardiac failure is still not fully understood. It seems that both quantitative as well as qualitative aspects of cardiac growth and development may be involved. In the first case, we may think of reaching the critical size of cardiac myocytes and of capillary domain, as discussed before. Qualitative aspects may involve changes in the expression of the genes that encode a number of key myocardial proteins resulting in the expression of their fetal isoforms, e.g. myosin heavy chains, actin, ß tropomyosin and creatine kinase, (3,4,19).

Acknowledgements

Our experiments described in this review have been supported by the Ontario Heart and Stroke Foundation and by the Medical Research Council of Canada.

References

1. Osler W. The Principles and Practice of Medicine: designed for the use of practitioners and students of medicine. 1892; Appleton, New York.
2. Meerson FZ. Circ Res 1969;25(Suppl II):1-163.
3. Katz AM. New England J Med 1990;322:100-110.
4. Katz AM. Cardiology 1990;77:346-356.
5. Boyd E. Origins of the Study of Human Growth. 1980; University of Oregon, Portland, OR.
6. Linzbach AJ, Akuamoa-Boateng E. Klin Wchschr 1973;51:156-163.
7. Olivetti G, Melissari M, Capasso JM, Anversa P. Circ Res 1991;68:1560-1568.
8. Rakusan K. In: Growth of the Heart in Health and Disease (ed.) R Zak. Raven Press, New York, 1984; pp. 131-164.
9. Simpson PC. Am J Cardiol 1988;62:13G-19G.
10. Nadal-Ginard B, Mahdavi V. J Clin Invest 1989;84:1693-1700.
11. Campbell SE, Rakusan K, Gerdes AM. Basic Res Cardiol 1989;84:247-258.
12. Rakusan K. In: Handbook of Human Growth and Developmental Biology. III/B (eds.) E Meisami, PS Timiras. CRC Press, Boca Raton LA, 1990; pp. 101-106.
13. Rakusan K, Turek, Z. Circ Res 1985;57:393-398.
14. Olivetti G, Anversa P, Loud AV. Circ Res 1980;46:503-512.
15. Batra S, Rakusan K. Am J Physiol (submitted for publication).
16. Rakusan K, Flanagan MF, Geva T, Southern J, Van Praagh R. Circulation (submitted for publication).
17. Rakusan K, Du Mesnil De Rochemont W, Braasch W, Tschopp H, Bing, RJ. Circ Res 1967;21:209-214.
18. Rakusan K, Korecky B. Can J Cardiol 1985;1:217-222.
19. Schneider MD, Parker TG. Circulation 1990;81:1443-1456.

Ontogenetic Development of Cardiac Inotropic Responsiveness

B. Ostádal, F. Kolár, *J. Skovránek, I. Ostádalová

Institute of Physiology, Czechoslovak Academy of Sciences,
**Kardiocentrum, University Hospital Motol,*
Prague, CZECHOSLOVAKIA

Introduction

Experimental data clearly indicate that significant ontogenetic differences exist in cardiovascular sensitivity to inotropic agents, both under "in vitro" and "in vivo" conditions (1). It seems likely that despite significant interspecies differences developmental changes may exist in all mammals depending on the degree of maturation of systems involved in the regulation of contractile function (2). This field of investigation was stimulated in recent years by an increasing clinical use of positive and negative inotropic drugs (e.g. catecholamines, calcium antagonists) during early phases of ontogenetic development, particularly during pregnancy in order to prevent premature labor or to treat acute intrapartum fetal distress (3, 4) or in pediatric cardiology (5, 6).

There are, however, many controversies concerning the mechanisms of developmental variations in cardiovascular responsiveness; the obvious discrepancies may be partly due to different experimental procedures used for studying different animal species (7). Nevertheless, it is worth noting that the ontogenetic conclusions are often based on observations from 2-3 developmental stages such as "fetal", "newborn" and "adult". It is therefore necessary to point out that the possible developmental changes between selected (and not exactly defined) points remain undetermined. The precise knowledge of individual ontogenetic periods critical for cardiac ontogeny (neonatal, suckling, weaning, adolescence, adulthood) is thus of crucial importance. Furthermore, the experimental data on the ontogenetic differences in cardiac contractile responsiveness were obtained exclusively on healthy individuals. This approach is, however, far from the clinical situation where interventions acting in the critical developmental periods may have serious consequences for a further maturation. The complications observed after the use of inotropic drugs during the early development (8, 9) may serve as warning examples.

Development of Myocardial Calcium Regulation and Inotropic Responsiveness

The contractile performance of the mammalian heart maturates during the postnatal period as evidenced by an increase in tension and pressure development and by changes in response to a variety of inotropic stimuli (Fig. 1) (10, 11). This has been largely attributed to a) an age-dependent change in α- and ß-myosin heavy chain expression resulting in a transition from low to high ATPase activity of myosin isoforms (12) and b) to the maturation of mechanisms controlling the intracellular Ca^{++} transients at the level of sarcoplasmic reticulum (SR) and/or sarcolemma. The mammalian myocardium is known to be dependent on both transsarcolemmal Ca^{++} influx and Ca^{++} release from the SR (2, 13). However, the relative contribution of the two mechanisms varies significantly during development (14). The SR of the newborn rat is not fully developed (15) and Ca^{++} induced release of Ca^{++} from SR is thus much less expressed as compared with adult animals. The contractility of neonatal myocardium depends, therefore, to a large extent on the entry of calcium across the sarcolemma via the Ca^{++} channels. During a further development the ability of SR to accumulate Ca^{++} increases and there is a progressive maturation of the Ca^{++} release from the SR (2). However, the exact time course of these ontogenetic changes in cellular calcium regulation is not known.

Many drugs that modify cardiac function do so by affecting myocardial calcium by changes in a) Ca^{++} fluxes, b) levels of

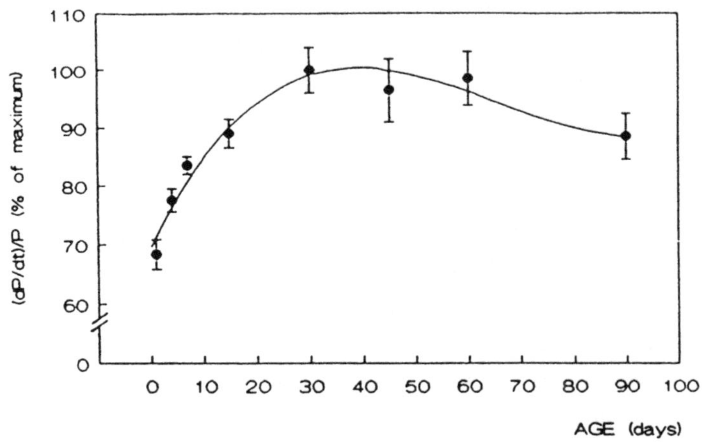

Fig. 1. Index of contractility (dP/dt/P); ontogenetic development; rat heart.

Ca^{++} at storage sites or c) Ca^{++} sensitivity of contractile proteins (16). They may, therefore, either increase cellular Ca^{++} and hence the force of contraction (positive inotropic drugs), or reduce cellular Ca^{++} loading and subsequently the developed tension (negative inotropic drugs). Since the regulatory mechanisms of intracellular calcium concentration change significantly during development, the type and degree of inotropic response is strictly age-dependent.

The developmental variations of contractile responses of the mammalian heart have been described for most positive and negative inotropic drugs; as has been mentioned above the results are highly controversial. For example, several investigators have shown that the positive inotropic effect of beta-mimetic catecholamines in newborn hearts was less than that observed in adult preparations (17, 18, 19). On the other hand, Nishioka et al (20) observed that the maximal inotropic effect in newborns was significantly greater than in adults: the sensitivity to isoproterenol was, however, similar in both age groups. The obvious discrepancies may be due to different experimental procedures as well as to different time-course of maturation of intracellular calcium regulation in different species.

Developmental changes in cardiac inotropic responsiveness should be, therefore, examined separately in each species.

For many years, only the positive inotropic drugs have been thought to be clinically useful. Recently, however, Ca^{++} antagonists (CA) which reduce rather specifically the Ca^{++} influx through the voltage dependent Ca^{++} channel in the cardiac and vascular smooth muscle have been found to have many clinical applications (16). Whereas the cardiovascular effects of CA have been studied extensively in adult animals, much less is known about their effect on the immature heart (21). In a short survey we would like to summarize some of our data concerning the effect of CA on the developing heart. Special attention was paid to the development of cardiac contractile response (particularly during the early phases of postnatal development) and to the hormonal interventions which accelerate or attenuate the maturation of systems involved in regulation of inotropic responsiveness.

Developmental Changes of Cardiac Sensitivity to Calcium Antagonists

In order to analyze the ontogenetic differences in cardiac sensitivity to CA, the mortality rate and the intensity of negative inotropic response of the isolated right ventricular myocardium (22) and isolated perfused hearts (23) were studied in rats during the postnatal development.

The mortality of verapamil treated animals was age-dependent. Whereas in 90 day old rats, the first death was observed only when a huge dose of $100.mg.kg^{-1}$ of verapamil had been injected, with decreasing age the mortality rate significantly increased. In newborn rats 25% mortality was still observed after $2.5\ mg.kg^{-1}$ of verapamil (Table 1).

The negative inotropic response of the rat isolated right ventricular myocardium (as judged from the amplitude of isotonic contractions) was significantly greater in 3 day old animals as compared with all older groups (15, 30, and 90 day old rats). Similar age related changes were observed in experiments using isolated perfused rat heart; this ontogenetic difference was found under the conditions of both constant perfusion pressure and constant

Table 1. Mortality rate after verapamil treatment (data from 22).

Verapamil s.c. ($mg.kg^{-1}$)	n	Mortality rate (%) rats - age (days)			
		3	15	30	90
0.5	10	0	0	0	0
1.0	10	0	0	0	0
2.5	8	25	0	0	0
5.0	10	60	0	0	0
7.5	10	90	80	0	0
10.0	8	100	75	0	0
25.0	10	100	100	0	0
50.0	10	100	100	90	0
100.0	8	100	100	100	12.5

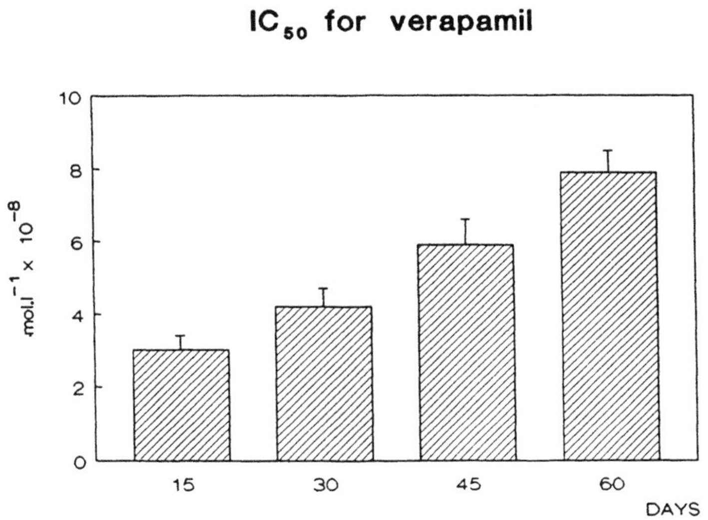

Fig. 2. Concentrations of verapamil that decreased cardiac contractility by 50%; postnatal ontogeny; rats; (data from 23).

Table 2. Concentration of verapamil that decreased cardiac contractility by 50% (IC_{50}).

Age group days	n	IC_{50} for verapamil (mol. $1.^{-1}$) mean ± S.E.M.
1	7	$(1.70 \pm 0.60) . 10^{-7}$
4	9	$(1.70 \pm 0.60) . 10^{-8}$
7	7	$(1.82 \pm 0.20) . 10^{-7}$

flow, and is thus independent of the state of coronary circulation.

In previous studies by Boucek et al (24) and Seguchi et al (25) on the rabbit myocardium, the conclusions concerning postnatal development of cardiac sensitivity to calcium antagonists were based on the comparison of newborn and adult hearts only, where the observed difference was prominent. It follows from our data, however, that the increased sensitivity to calcium channel blockers manifests itself throughout maturation: 30-day-old rats still showed a significantly lower value of IC_{50} for verapamil than did the adult hearts (Fig. 2).

As obvious from Fig. 1, neonatal period is of crucial importance for the further development of myocardial contractile function. Unfortunately, the time-limitation of this critical stage of development is in the cardiological literature often very uncertain and rather broad (it varies from 0 to 20 days of postnatal life in rabbits, from 0 to 30 days in rats). Recently, we have shown (26) that the inotropic effect of low extracellular sodium in isolated perfused rat heart develops dramatically during the first week of life. The shape of the curve of developed force changed day by day and only on day 7 did it resemble the triphasic response observed in adult rats (27).

Similar impressive neonatal changes can be observed in experiments with CA (26). The most marked negative inotropic response of the isolated perfused rat heart was observed not on day 1 of postnatal life as expected but only on day 4 (Table 2). Later on the negative inotropic response decreased in accordance with the previously published results (27). A more detailed analysis of the first 3 days of life will be the aim of further investigations.

The ontogenetic differences in the cardiac sensitivity to CA are obviously due to the different role played by distinct membrane systems (i.e. plasma membrane and SR) in the regulation of cardiac contractility during postnatal development. Many experimental data suggest that the internal cycle of calcium, which is dependent on accumulating and sequestering abilities of SR, is not well developed in the immature heart (11). In this respect, the immature mammalian heart is like the heart of poikilothermic animals, in which the extracellular space is the only source of calcium directly activating the contractile apparatus; transsarcolemmal fluxes are thus the major determinants of contractile force. Under such developmental conditions it is not surprising that CA may have a detrimental effect on the neonatal heart.

It seems that the ontogenetic differences in the effect of CA are not the result of changes at the level of receptors because both the density and the affinity of nitrendipine-binding sites attain adult values in the rat myocardium already during the first week of postnatal life. Similarly, in the rabbit heart, the binding characteristics did not differ in neonatal and adult age (28, 29).

It may be, therefore, concluded that a) the sensitivity of cardiac contractile function to CA declines gradually during postnatal ontogeny and b) the higher sensitivity of the developing heart to calcium channel blockade is a consequence of a higher functional dependence of the immature myocardium on transsarcolemmal calcium influx.

Thyroid Control of Early Postnatal Development of Cardiac Inotropic Responsiveness

Theoretically, the contribution of individual systems involved in regulation of cardiac contractility and inotropic responsiveness may be affected by interventions which accelerate or attenuate their maturation. A typical example is the thyroid status: thyroid hormones could modulate the developing heart at least a) via altering the amount and properties of different contractile proteins (30, 31); b) via controlling intracellular calcium transient at the level of SR (32, 33) and/or sarcolemma (34, 35) and c) via increased hemodynamic load (36).

Normal ontogenetic development of cardiac contractile function, myosin isoforms expression and Ca^{++} handling resemble changes induced by excess thyroid hormone in adult euthyroid or hypothyroid animals (31, 37). Since in rats thyroid hormones appear in fetal plasma on day 18 (38) and their levels reach a maximum during the third postnatal week (39), they could be expected to serve as hormonal modulators of the development of contractile properties of the cardiac muscle, in particular in the maturation of Ca^{++} transport mechanisms of cardiac membranes, and thus to change cardiac inotropic responsiveness. We therefore examined the influence of an altered thyroid state 1) on the basal contractile performance, 2) on the inotropic responsiveness to a sarcolemmal Ca^{++} channel blocker (verapamil) and to the SR inhibitor (ryanodine) in three week old rats (40).

Neonatal rats were rendered hyperthyroid by daily subcutaneous injection of L-triiodothyronine (T_3, Serva) in a dose of 10 $ug.100g^{-1}$ body weight starting from the 2nd day postpartum and continuing for 3 weeks. Hypothyroidism was induced by the inclusion of 0.03% 6-n-propyl-2-thiouracil (PTU) in the drinking water to mothers also from day 2 to day 21.

Both hypo- and hyperthyroid animals exhibited significant growth retardation; body weight reached approximately 50% of the euthyroid level. Thyroxine-induced cardiomegaly was obvious from the values of the relative heart weight; it was more pronounced in the right ventricle. The hypothyroid rats were characterized by an almost three-fold reduction in total heart weight as compared with the euthyroid group.

The isolated perfused hyperthyroid hearts exhibited markedly higher values of the left ventricular DP and $+(dP/dT)_{max}$ which could be in part attributed to the relative enlargement of the ventricle. Hypothyroidism was associated with the opposite changes (Fig. 3). The addition of verapamil into the perfusion medium caused concentration-dependent negative inotropic effect in all groups of hearts studied. In the hypothyroid group the relative depression of

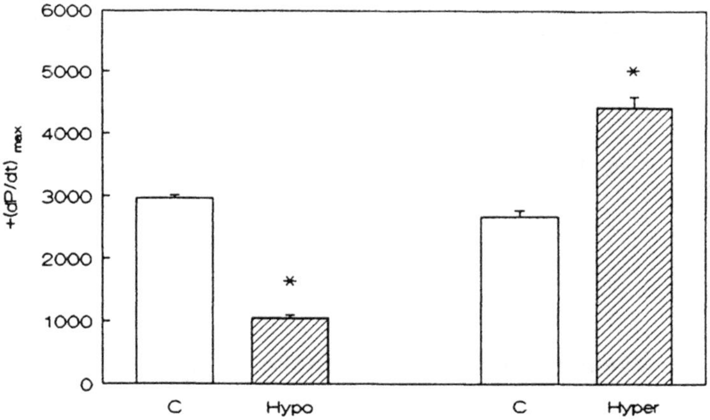

Fig. 3. Effect of altered thyroid state on contractile parameter of the Langendorff-perfused rat hearts; $+(dP/dt)_{max}$ - maximum rate of pressure development; C - controls; Hypo - hypothyroid rats; Hyper - hyperthyroid rats; statistical significance - $p < 0.05$; (data from 40).

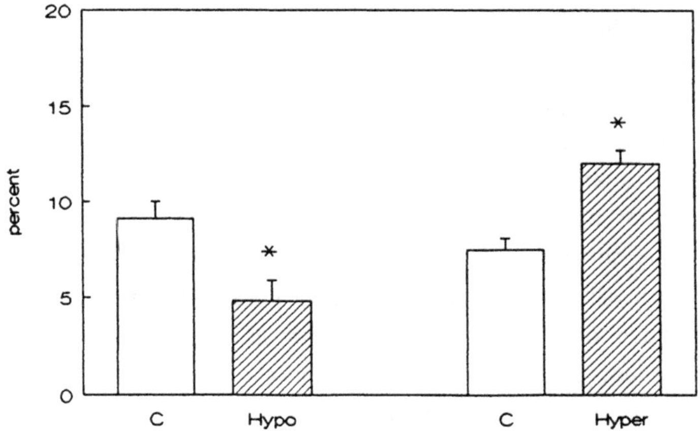

Fig. 4. Concentrations of verapamil that decreased cardiac contractility by 50% (IC_{50}); C - controls; Hypo - Hypothyroid rats; Hyper - hyperthyroid rats; statistical significance - $p < 0.05$; (data from 40).

EFFECT OF THYROID STATUS

	HYPO	HYPER
negative inotropic response		
verapamil	↑	↓
ryanodine	↓	↑
oxalate supported Ca^{++} uptake	↓	↑

Fig. 5. (data from 40).

$+(dP/dT)_{max}$ was more prominent than in the euthyroid group at any verapamil concentration. On the other hand, in the hyperthyroid hearts the effect of verapamil was less expressed than in the euthyroid ones. Thus, compared to the euthyroid state, hypo-and hyperthyroidism were associated with increased and decreased sensitivity to the negative inotropic response to verapamil, respectively, as evident from significant shifts in the IC_{50} values (Fig. 4). The contribution of the transsarcolemmal Ca^{++} fluxes to the control of cytosolic Ca^{++} transients thus seems to be diminished in hyperthyroid and increased in hypothyroid rats as compared with euthyroid animals of the same age.

The alterations in the thyroid state also led to changes in the negative inotropic response to ryanodine added into the perfusion medium in a concentration of 10^{-6}M. In the hypothyroid hearts the $-(dP/dt)_{max}$ was inhibited by this drug significantly less than in the euthyroid ones, while the depression of DP and $+(dP/dt)_{max}$ did not differ between these groups. On the contrary, in hyperthyroid hearts, the negative inotropic response to ryanodine was more pronounced than in euthyroid hearts as evident from a significantly higher inhibition of all three contractile parameters. The different responsiveness of the hypo- and hyperthyroid immature myocardium to the SR Ca^{++} release channel inhibitor ryanodine supports the view that the contribution of the SR mediated Ca^{++} cycling to the control of cardiac contractile function increases after excess thyroid hormone treatment, thus facilitating the normal developmental process.

It may be concluded that the cardiac contractile function and inotropic responsiveness during early postnatal development seem to be markedly affected by altered thyroid state. In hyperthyroidism, the postnatal maturation of the SR - mediated Ca^{++} handling is accelerated and the heart becomes less dependent on transsarcolemmal Ca^{++} fluxes; hypothyroidism leads to the opposite changes (Fig. 5).

Conclusions

Experimental data clearly indicate that significant ontogenetic differences exist in the cardiovascular sensitivity to inotropic agents both under "in vitro" and "in vivo" conditions. It seems likely that despite significant interspecies differences, developmental changes may exist in all mammals depending on the degree of maturation of the systems involved in the regulation of contractile function. Accordingly, significant age-related changes were observed in cardiac sensitivity to calcium antagonists (CA). The mortality of CA treated rats decreased with increasing age of the animals. Similarly, the sensitivity of the isolated heart to the negative inotropic effect of CA declines gradually during development. Cardiac responsiveness to CA may be markedly influenced by neonatal thyroid status: hyperthyroidism decreased, whereas hypothyroidism increased the negative inotropic response to CA. The higher sensitivity of the immature myocardium to CA is obviously a result of a greater functional dependence on transsarcolemmal calcium influx into the cells via voltage-operated calcium channels. These results point to the possible negative consequences of the clinical use of inotropic agents during early phases of ontogenetic development.

References

1. Driscoll D. J Clin Perinatol 1987;14:931-948.

2. Fabiato A, Fabiato F. Ann NY Acad Sci 1978;307:491-522.
3. Benedetti T, J Clin Perinatol 1986;13:843-852.
4. Harake B, Glibert RD, Ashwal S, Power GG. Am J Obstet Gynecol 1987;157:1003-1008.
5. Sapire DW, O'Riordan AC, Black IFS. Am J Cardiol 1981;48:1091-1097.
6. Cho C, Pruitt AW. Am J Dis Child 1986;140:360-366.
7. Ostádal B, Beamish RE, Barwinsky J, Dhalla NS. J Appl Cardiol 1989;4:467-486.
8. Abinader E, Borowitz Z, Berger A, Helv Pediatr Acta 1981;36:451-455.
9. Wischnik A, Hotzinger B, Wischnik B, Schroll A, Trenkwalder V, Wiedenback A. Arzneim Forsch 1984;34:684-687.
10. Artman M, Kithas PA, Wike JS, Strada SJ. Am J Physiol 1988;255:H335-H342.
11. Nakanishi T, Seguchi M, Takao A. Experientia 1988;44:936-944.
12. Swynghedauw B. Physiol Rev 1986;66:710-771.
13. Langer GA. In: Calcium and the Heart (eds.) GA Langer. Raven Press Ltd, New York, 1990;355-378.
14. Tanaka H, Shigenobu K. J Mol Cell Cardiol 1989;21:1305-1313.
15. Schiebler TH, Wolf HH. Zellforsch 1966;69:22-40.
16. Akera T. In: Calcium and the Heart (eds.) GA Langer. Raven Press Ltd, New York, 1990;299-331.
17. Park MK, Sheridan PH, Morgan WF, Beck N. Devol Pharmacol Ther 1980;1:70-82.
18. Rockson SG, Homcy CJ, Quinn P, Manders WT, Haber E, Vatner S. J Clin Invest 1981;67:319-327.
19. Driscoll DJ, Fukushige J, Hartley CJ, Lewis RM, Entmann ML. Devel Pharmacol Ther 1981;2:91-103.
20. Nishioka K, Nakanishi T, George BL, Jarmakani JM. J Mol Cell Cardiol 1981;13:511-520.
21. Ostádal B, Skovránek J, Kolár F, Janatová T, Krause EG, Ostadálová I. Biomed Biochim Acta 1987;46:S522-S526.
22. Skovránek J, Ostádal B, Pelouch V, Procházka J. Pediatr Cardiol 1986;7:25-29.

23. Kolár F, Ostádal B, Papousek F. Basic Res Cardiol 1990;85:429-434.
24. Boucek RJ, Shelton ME, Artman M, Mushlin PS, Starnes VA, Olson RD. Pediatr Res 1984;18:948-952.
25. Seguchi M, Jarmakani JM, George BL, Harding JA. Pediatr Res 1986;20:838-842.
26. Ostádalová I, Kolár F, Rohlicek J, Rohlicek V, Procházka J, Ostádal B. J Mol Cell Cardiol 1992; (in press).
27. Kolár F, Cole WC, Ostádal B, Dhalla NS. Am J Physiol 1990;259:H712-H719.
28. Renaud FJ, Kazazoglou T, Schmid A, Romey G, Lazdunski M. Eur J Biochem 1984;139:673-681.
29. Boucek RJ, Shelton ME, Artman M, Landon E. Basic Res Cardiol 1985;80:316-325.
30. Chizzonite RA, Zak R. J Biol Chem 1984;259:12628-12632.
31. Lompre AM, Nadal-Ginard B, Mahdavi V. J Biol Chem 1984;259:6437-6446.
32. Limas CJ. Am J Physiol 1978;235:H745-H751.
33. Takacs IE, Nosztray K, Szabo J, Szentmiklos AJ, Cseppento A, Szegi J. Gen Physiol Biophys 1985;4:271-278.
34. Kim D, Smith TW, Marsh JD. J Clin Invest 1987;80:88-94.
35. Daly MJ, Seppet EK, Vetter R, Dhalla NS. In: Subcellular Basis of contractile Failure (eds.) B Korecky, NS Dhalla. Kluwer Academic Publishers, Boston; 1990;173-191.
36. Strauer BE, Schulze W. Bas Res Cardiol 1976;71:624-644.
37. Marriott ML, McNeil JH. Can J Physiol Pharmacol 1983;61:1382-1390.
38. Pic P, Bouguin J-P. J Dev Physiol 1985;7:207-214.
39. Vigouroux E. Acta Endocrinol 1976;83:752-762.
40. Kolár F, Seppet EK, Vetter R, Procházka J, Grünermel J, Zilmer K, Ostádal B. Pflügers Arch 1991; (in press).

B. REGULATION OF CARDIAC CHANNELS

Activation Mechanisms of Cardiac Muscarinic K Channels

H. Heidbüchel and E. Carmeliet

Laboratory for Electrophysiology, University of Leuven,
Gasthuisberg O & N, Herestraat 49, B-3000 Leuven, BELGIUM

Introduction

Muscarinic K^+ channels form a distinct class of cardiac channels, with conductance and kinetic properties that distinguish them from other types of cardiac K^+ channels. They are called "muscarinic K^+ channels" because they were first identified as the channels responsible for the acetylcholine-induced increase of K^+ conductance in cardiac tissue (1). The induction of a K^+ conductance increase by acetylcholine (ACh) can be completely blocked by atropine, implying that this action of ACh is mediated by muscarinic receptors. The activation of an inwardly rectifying K^+ current is the most prominent cellular response after vagal stimulation or the application of exogeneous ACh in atrial, nodal and Purkinje cells (1-5). The acetylcholine-induced K^+ current ($i_{K(ACh)}$) results in hyperpolarization, decreased rate of spontaneous diastolic depolarization and shortened action potential duration. The K^+ conductance increase contributes to the negative chronotropic, negative inotropic and negative dromotropic effect of vagal stimulation. Four other cellular effector pathways further contribute to the cardiac effects of parasympathetic stimulation (for an overview see (6)): 1] antagonism of the production and of the effects of cAMP; 2] stimulation of phosphoinositide turnover; 3] direct reduction of the pacemaker current i_f (7); and 4] stimulation of the cellular cGMP content. $IK_{(ACh)}$ generally is not present in mammalian ventricular cells, although exceptions have been reported for the ferret (8). In 1985, it was recognized that muscarinic activation of the channels required intracellular GTP due to the involvement of a guanine-nucleotide binding protein (G protein) (9-10). It is the aim of this text to briefly review the basic mechanism of G protein-mediated K^+ channel activation by muscarinic stimulation and to highlight some important variations on the general theme.

Muscarinic K^+ Channel Activation: Basic Theme

The Muscarinic Receptor

Acetylcholine receptors can be divided in two largely different categories: the nicotinic receptor (present in the neuroeffector junction of skeletal muscle and in the central nervous system) and the predominantly visceral muscarinic receptors. The muscarinic receptors belong to a group of receptors that are all coupled to GTP-binding proteins. The group includes, among others, the adrenergic, dopaminergic, serotonergic receptors, and the photosensitive pigment rhodopsin. The receptors of this group share structural similarities: they all contain about 440 amino acids of which 22% to 71% are identical (11). The predicted ternary configuration includes 7 hydrophobic regions, which probably form transmembrane spans. First sequence analysis of the muscarinic receptors from porcine brain and heart was reported in 1986 (12-13). Five subtypes of the muscarinic receptor have been identified, called m1 till m5, all coded by different genes. In cardiac cells, only the m2-receptor seems to be present (14-15). Although the structure of the different muscarinic receptors is highly homologous (60% of the amino acids being identical), there is some specificity for the effector systems, albeit not exclusive. The m1, m3 and m5 receptors, and the m2 and m4 receptors resemble each other most. The first series is strongly coupled with phosphatidylinositol turnover; the m2 and m4 receptors strongly inhibit adenylyl cyclase but they can also weakly stimulate phosphatidylinositol turnover (16-18).

The G Protein

G proteins are heterotrimeric proteins consisting of a guanine-nucleotide binding α-subunit (which is specific for the different types of G proteins) and a $\beta\gamma$ dimer (thought to be shared by many G proteins). Binding of agonist to the muscarinic receptor leads to a GDP-to-GTP exchange on the G protein α-subunit (Figure 1). The subsequent dissociation of the α-GTP from the $\beta\gamma$ subunits constitutes the activation of the G protein. Intracellular Mg^{2+} is essential for binding of the nucleotide and subsequent dissociation of the G protein (19-21). GTP hydrolysis by the intrinsic GTPase activity of the α-subunit, followed by reassociation of α-GDP with $\beta\gamma$ ends the activation cycle (20-22). In the absence of agonists, the GTPase activity is at least ten times greater than the rate of guanine nucleotide exchange; therefore, very little G protein is in the active α-GTP configuration under these conditions.

The G protein which is responsible for the activation of muscarinic K^+ channels is of the same type as the G protein which mediates muscarinic inhibition of adenylyl cyclase. It is generally denoted as G_i (from "inhibitory"), but also the name G_K (from "K^+ channel activating") is used. The α_i-subunit itself exists as three different subtypes (α_i-1 to α_i-3): all are equally effective in activating the K^+ channels (23-24). Also another type of G protein, G_o, which is abundant in brain, can open muscarinic K^+ channels (25). In cardiac tissues of both rat and human, the presence of α_o has been shown (26). The G protein, responsible for the activation of PLC (denoted as G_p) is believed to be different (27), although G_i/G_K-type G proteins are able to (weakly) activate PLC (17). The involvement of different G proteins is important from a pharmacological and experimental viewpoint, since G_i/G_K can be covalently modified and inhibited by pertussis toxin catalyzed ADP-ribosylation, whereas G_p is pertussis toxin-insensitive (27).

The Muscarinic K Channel

The channels responsible for the ACh-induced current in heart tissue, share some characteristics with other types of cardiac K^+

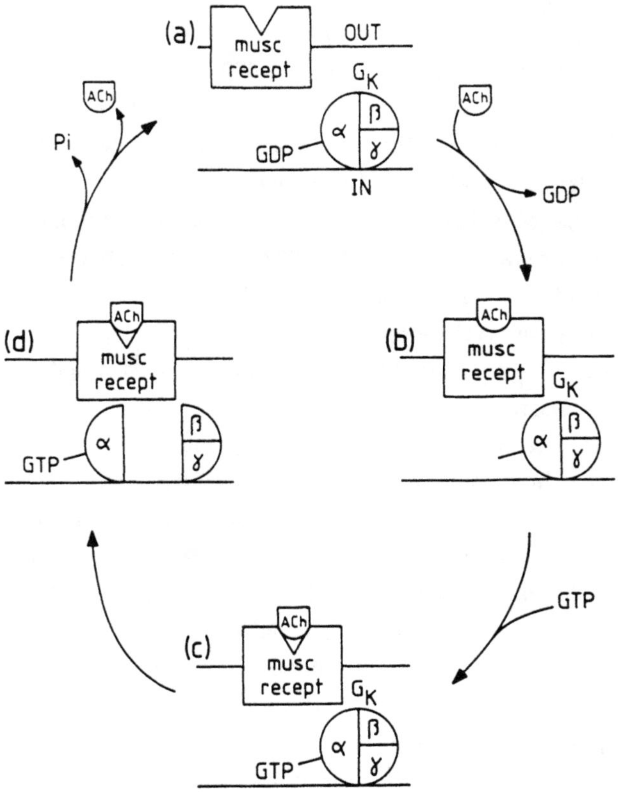

Fig. 1. Activation/deactivation cycle of the heterotrimeric G protein G_K. In the absence of agonist, the GDP which is bound to the α-subunit, exchanges very slowly with free intracellular GTP. The G protein is mainly in the undissociated, inactive configuration (a). ACh binding to the muscarinic receptor facilitates exchange of the of the bound GDP with GTP (b & c), resulting in activation of the G protein by dissociation of the α-GTP from the Bγ subunits (d). Hydrolysis of GTP and reassociation of the three subunits ends the activation cycle (d to a).

channels, e.g. their high selectivity for K^+ ions and the fact that they pass more current in the inward direction than to the exterior of the cell (inward rectification). Other properties however, allow to distinguish them clearly as a separate class. The single channel conductance is dependent on the external K^+ concentration (K_e) and is of the order of 40 pS for $K_e \approx 140$ mEq/1; openings are of short duration and are grouped in clusters, called bursts. The mean open time is 1 to 3 ms in all species: rabbit (1, 28), guinea pig (23, 29-31), chick embryo (25, 32) and human (33). The channels sporadically open in the absence of ACh (basal activity) but the frequency of opening is highly increased by the extracellular application of agonist. GTP at the intracellular (cytoplasmic) side of the membrane is essential for this agonist-induced activation, demonstrating that transmembrane signalling involves G proteins. The mean open time of the short openings is identical in the basal state or after activation, and is the same for all known activation modes of the channels. The only exception is a 5-fold increase of the mean open time by a cAMP-dependent protein kinase, an effect which is reversible by alkaline phosphatase (34). The closed times within bursts are shorter than the time-periods between bursts: the shorter mean closed time is from the order of the mean open times (±1ms); the longer one depends on the concentration of agonist, since ACh application will result in more bursts with shorter interburst periods. Protein kinase A also had a minor effect on the (short type) closed time, which was doubled (34).

Muscarinic K^+ Channel Activation: Variations

G protein-mediated channel activation: α or $\beta\gamma$?

Although dissociation of the GTP-bound G_γ protein is a prerequisite for its signal transducing activity, it is far from clear how much the free α-subunit or the $\beta\gamma$ dimer contribute to the final responses. K^+ channel activation was first shown to be mediated by α_i subunits (35), but soon it was reported that also exogenously applied $\beta\gamma$ subunits could activate (25, 30-32). The activation by α is probably direct, i.e. without further

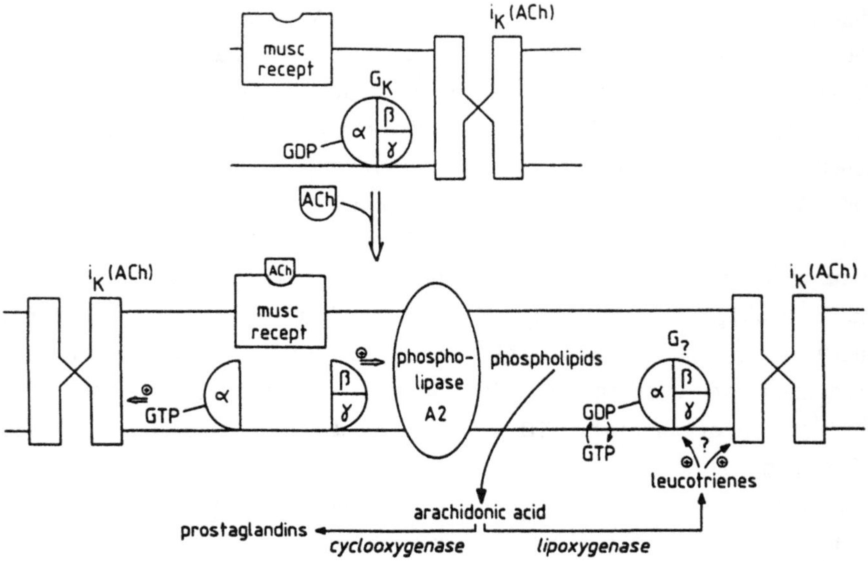

Fig. 2. ACh-induced K^+ channel activation can be mediated by both α-GTP and Bγ subunits of G_K. The GTP-bound α subunit is thought to directly induce K^+ channel openings. The activation by B is probably mediated by phospholipase A_2 stimulation, liberation of arachidonic acid, and formation of leucotrienes. The exact mechanism by which channels are finally opened remains unclear. See text for further discussion.

(cytoplasmic) intermediates (28). In contrast, the B subunits are thought to open the channels by activation of a membrane-bound phospholipase A_2 : the liberated arachidonic acid (AA) is metabolized by lipooxygenase to leucotrienes which in turn will activate the K^+ channels (Figure 2) (36-37). It was reported that this AA-mediated activation required GTP to activate the muscarinic K^+ channels in guinea pig atrial cells, suggesting stimulation by AA or its metabolites of guanine-nucleotide exchange on G_K (36). In rat atrial cells however, AA could stimulate in the absence of GTP (37). Remarkably, another group could not confirm the Bγ stimulatory action using Bγ dimers from four different origins, and reported that these subunits would rather have an inhibitory action by their mass-action reversal of α-from-Bγ dissociation (30,35); it was postulated that the $\beta\gamma$ subunits serve to improve the signal-to-noise ratio of muscarinic signal transduction since they inhibit basal activity more than agonist-stimulated activity (38). Thus, the definite conclusion about $\beta\gamma$'s stimulating capacity and the exact activation sequence (involving lipid-metabolites) remains to be elucidated. Anyway, the activation by α-GTP seems to be the physiologically relevant one: 1] inhibition of the $\beta\gamma$ effects (by antibodies against PLA_2 or by lipoxygenase inhibitors) does not disturb ACh-mediated channel activation (37, 39-40), and 2] application of a monoclonal antibody to the α-subunit of G_i/G_K, fully blocks muscarininc activation (41).

Other Neurotransmittors Activate the Same Channels

Not only muscarinic agonists activate the "ACh-dependent" K^+ channels. Also adenosine (via P_1-type purinergic receptors (33, 42-44), extracellular ATP (via P_2-type receptors (45)), somatostatin (via another specific receptor (46)), and alpha-adrenergic agonists (via α_1-receptors (47)) open channels with the same characteristics. The P_1-, P_2- and somatostatin receptors are coupled to G_K, and open the same population K^+ channels as do muscarinic agonists. K^+ channel opening by α_1-agonists is not mediated by G_K-activation, but includes phospholipase A_2 stimulation, AA liberation, and the intermediate formation of leucotrienes by 5-lipoxygenase. Therefore, AA-metabolites may be common intracellular second messengers to muscarinic channel activation (48-49). We already mentioned that pertussis toxin (PTX) catalyzes a NAD-dependent covalent modification of G_i/G_K which blocks the normal transduction cascade (29). Interestingly, not only PTX is able to uncouple the muscarinic receptor from the G protein. A small effect has been described for intracellularly applied heparin (50), and for the complex of p21-ras and GAP (GTPase activating protein) (51). In how far these experimental uncoupling mechanisms have correlated for physiological or pathophysiological control of muscarinic K^+ channel activity in vivo, is unknown at present.

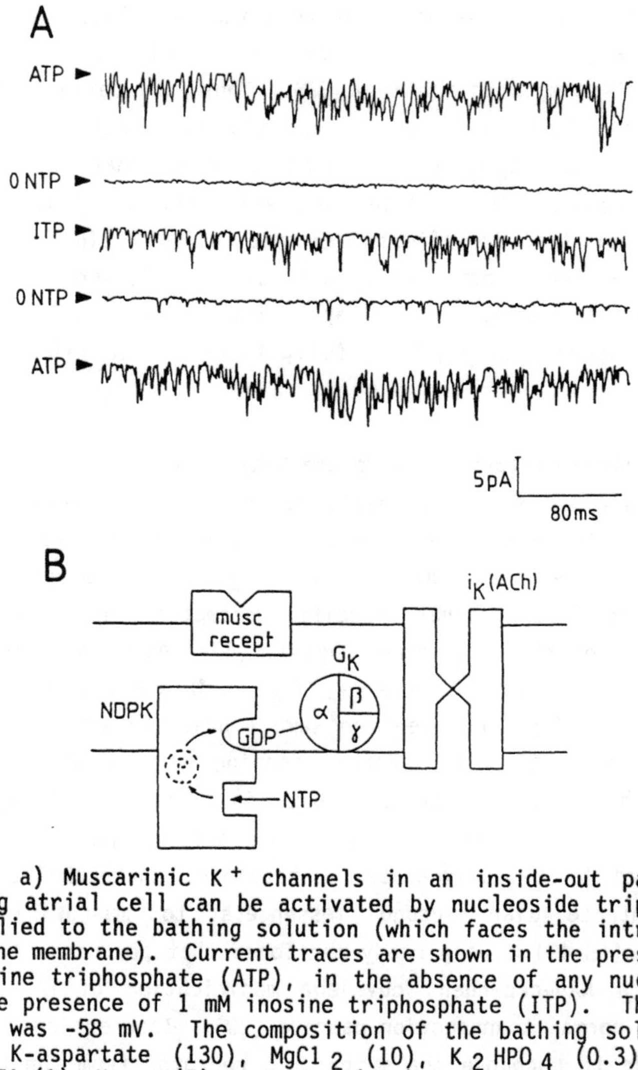

Fig. 3. a) Muscarinic K^+ channels in an inside-out patch of a guinea pig atrial cell can be activated by nucleoside triphosphates (NTP) applied to the bathing solution (which faces the intracellular side of the membrane). Current traces are shown in the presence of 1 mM adenosine triphosphate (ATP), in the absence of any nucleotides, and in the presence of 1 mM inosine triphosphate (ITP). The holding potential was -58 mV. The composition of the bathing solution was (in mM): K-aspartate (130), $MgCl_2$ (10), K_2HPO_4 (0.3), glucose (5.5), EGTA (1), Hepes (5) at pH 7.3 (KOH). The patch-clamp pipettes contained KCl (140), $CaCl_2$ (1.8), $MgCl_2$ (1), Hepes (5) at pH 7.3, and no agonist. b) Schematic drawing of the NTP-mediated activation of muscarinic K^+ channels. A membrane-bound nucleoside diphosphate kinase (NDKP) is associated with the triad muscarinic receptor -G_K-K^+ channel. The enzyme can transfer -phosphate groups from NTPs to NDPs. In the absence of exogenous guanine-nucleotides, G_K-bound GDP will fit into the acceptor-site on the NDPK, and will be converted into GTP. This will activate the G protein and result in channel openings.

Activation of Muscarinic K^+ Channels by a Membrane-Bound Nucleoside Diphosphate Kinase (NDPK).

We and others have shown that guinea pig, rabbit and frog atrial muscarinic K^+ channels in cell-free pieces of membrane confined to the tip of a microelectrode (patches) can be activated in the absence of agonist and guanine-nucleotides, provided that cytolsolic ATP and Mg^{2+} are present (52-54). Not only ATP can activate the channels under these conditions but also other nucleoside triphosphates (NTP) result in channel openings, as is illustrated for inosine triphosphate in Figure 3A. This NTP-dependent activation of the channels is compatible with a direct phosphorylation of G_K-bound GDP by a membrane associated NDPK, which transphosphorylates the gamma-phosphate of any NTP to G_K-bound GDP. As a result, GDP is transformed to GTP directly on the G protein: the G protein is activated, dissociates and leads to opening of the channels. A model of the proposed interaction of NDPK with the transmembrane transduction pathway, is shown in Figure 3B. We identified the presence of NDPK in frog and guinea pig atrial membranes by demonstrating and characterizing its enzymatic activity. Membrane vesicles were incubated with different combinations of tritium-labelled NDPs and NTPs in order to study the transformation of these nucleotides. The order of efficiency as a phosphate-donor (using GDP as acceptor) was ATP > GTP ≅ ITP > UTP > CTP, which parallels the ability of these different nucleotides to activate the channels in isolated patches. These results support the concept of the association of a membrane-bound NDPK to the receptor -G_K - K^+ channel cascade.

Interestingly, submicromolar concentrations of GDP or GTP blocked the NTP-dependent activation of the muscarinic K^+ channels, by competing with the G_K-bound GDP for the NDPK. Therefore, under normal physiological conditions and in the absence of agonist, the high intracellular ATP-concentrations will not activate the channels because intracellular GDP and GTP will block this activation. However, sporadic transphosphorylation of G_K-bound GDP may be an alternative mechanism for non-stimulated GDP-to-GTP exchange to account for the basal activity of the channels in the absence of agonists. We have observed that in 9 out of 19 inside-out experiments, ACh-dependent stimulation of muscarinic K^+ channels was possible in the presence of 4 mM ATP + 0.1 mM GDP, in contrast with the complete absence of channel activation in the absence of ACh under these conditions. This indicates that even in continuously perfused patches, NDPK is able to synthesize enough local GTP as cofactor for agonist-induced channel activation. Therefore, the main functional role of NDPK under physiological conditions would be to provide a local supply of GTP (using GDP and ATP) in the immediate vicinity of the G protein, thus maintaining a high local GTP/GDP ratio and ensuring an adequate receptor-mediated activation of the muscarinic K^+ channels.

Conclusion

It is clear that the regulation of muscarinic K^+ channels is much more complex than the simple pathway "receptor - G_K - K^+ channel" would suggest at first glance. Not only ACh, but also adenosine, ATP, somatostatin and α_1-agonists, acting via specific receptors, are able to open the channels via the intermediate activation of different G proteins. The α-subunits of G_i-1, G_i2, G_i-3 and G_o (all denoted as "G_K") directly open the channels. Also $\beta\gamma$ -subunits may activate the channels via stimulation of phospholipase A_2 (leading to the formation of arachidonic acid and leucotrienes), but in how far this pathway is physiologically relevant remains unknown. Sympathetic stimulation can modulate the open probability of the muscarinic K^+ channels via protein kinase A, and the complex of p21-ras and GAP-protein has been shown to uncouple signal transduction in vitro. Membrane-bound nucleoside diphosphate kinase can directly activate G_K in the absence of G-nucleotides, and may serve as a local supply of GTP in vivo, ensuring efficient receptor-mediated activation of the muscarinic K^+ channels.

References

1. Sakmann B, Noma A, Trautwein W. Nature 1983;303:250-253.
2. Trautwein W, Dudel J. Pflügers Arch 1958;266:324-334.

3. Garnier D, Nargeot J, Ojeda C, Rougier O. J Physiol Lond 1978;274:381-396.
4. Carmeliet E, Mubagwa K. J Physiol Lond 1986;371:201-217.
5. Heidbüchel H, Vereecke J, Carmeliet E. J Mol Cell Cardiol 1987;19:1207-1219.
6. Hartzell HC. Prog Biophys Mol Biol 1988;52:165-247.
7. Yatani A, Okabe K, Codina J, Birnbaumer L, Brown AM. Biophys J 1990;57:291a.
8. Boyett MR, Kirby MS, Orchard CH, Roberts A. J Physiol Lond 1988;404:613-635.
9. Pfaffinger PJ, Martin JM, Hunter DD, Nathanson NM, Hille B. Nature 1985;317:536-538.
10. Breitwieser GE, Szabo G. Nature 1985;317:538-540.
11. O'Dowd BF, Lefkowitz RJ, Caron MG. Annu Rev Neurosci 1989;12:67-83.
12. Kubo T, Fukuda K, Mikami A, Maeda A, Takahashi H, Mishina M, Haga T, Haga K, Ichiyama A, Kangawa K, Kojima M, Matsuo H, Hirose T, Numa S. Nature 1986;323:411-416.
13. Kubo T, Maeda A, Sugimoto K, Akiba I, Mikami A, Takahashi H, Haga T, Haga K, Ichiyama A, Kangawa K, Matsuo H, Hirose T, Numa S. FEBS Lett 1986;209:367-372.
14. Peralta EG, Ashkenazi A, Winslow JW, Smith DH, Ramachandran J, Capon DJ. EMBO J 1987;6:3923-3929.
15. Maeda A, Kubo T, Mishina M, Numa S. FEBS Lett 1988;239:339-342.
16. Peralta EG, Ashkenazi A, Winslow JW, Ramachandran J, Capon DJ. Nature 1988;334:434-437.
17. Lechleiter J, Peralta E, Clapham D. TIPS Suppl Subtypes of muscarinic receptors 1989;IV:34-38.
18. Fukuda K, Kubo T, Maeda A, Akiba I, Bujo H, Nakai J, Mishina M, Higashida H, Neher E, Marty A, Numa S. TIPS Suppl Subtypes of muscarinic receptors 1989;IV:4-10.
19. Kurachi Y, Nakajima T, Sugimoto T. Pflügers Arch 1986;407:572-574.
20. Gilman AG. Annu Rev Biochem 1987;56:615-649.
21. Horie M, Irisawa H. J Physiol Lond 1989;408:313-332.
22. Brown AM, Birnbaumer L. Am J Physiol 1988;254:H401-410.

23. Yatani A, Mattera R, Codina J, Graf R, Okabe K, Padrell E, Iyengar R, Brown AM, Birnbaumer L. Nature 1988;336:680-682.
24. Mattera R. Yatani A, Kirsch GE, Graf R, Okabe K, Olate J, Codina J, Brown AM, Birnbaumer L. J Biol Chem 1989;264:465-471.
25. Logothetis DE, Kim DH, Northup JK, Neer EJ, Clapham DE. Proc Natl Acad Sci USA. 1988;85:5814-5818.
26. Urasawa K, Murakami T, Yasuda H. J Mol Cell Cardiol 1988;20(Suppl I):S49.
27. Masters SB, Martin MW, Harden TK, Brown JH. Biochem J 1985;227:933-937.
28. Soejima M, Noma A. Pflügers Arch 1984;400:424-431.
29. Kurachi Y, Nakajima T, Sugimoto T. Am J Physiol 1986;251:H681-H684.
30. Codina J, Yatani A, Grenet D, Brown AM, Birnbaumer L. Science 1987;236:442-445.
31. Cerbai E, Klockner U, Isenberg G. Science 1988;240:1782-1783.
32. Logothetis DE, Kurachi Y, Galper J, Neer EJ, Clapham DE. Nature 1987;325:321-326.
33. Heidbüchel H, Vereecke J, Carmeliet E. Circ Res 1990;66:1277-1286.
34. Kim D. Circ Res 1990;67:1292-1298.
35. Yatani A, Codina J, Brown AM, Birnbaumer L. Science 1987;235:207-211.
36. Kurachi Y, Ito H, Sugimoto T, Shimizu T, Miki I, Ui M. Nature 1989;337:555-557.
37. Kim D, Lewis DL, Graziadei L, Neer EJ, Bar Sagi D, Clapham DE. Nature 1989;337:557-560.
38. Okabe K, Yatani A, Evans T, Ho YK, Codina J, Birnbaumer L, Brown AM. J Biol Chem 1990;265:12854-12858.
39. Kurachi Y, Ito H, Sugimoto T, Katada, Ui M. Pflügers Arch 1989;413:325-327.
40. Bourne HR. Nature 1989;337:504.
41. Yatani A, Hamm H, Codina J, Mazzoni MR, Birnbaumer L, Brown AM. Science 1988;241:828-831.
42. Belardinelli L, Isenberg G. Am J Physiol 1983;244:H734-H737.
43. Hutter OF, Rankin AC. J Physiol Lond 1984;353:111-125.

44. Kurachi Y, Nakajima T, Sugimoto T. Pflügers Arch 1986;407:264-274.
45. Friel DD, Bean BP. Pflügers Arch 1990;415:651-657.
46. Lewis DL, Clapham DE. Pflügers Arch 1989;414:492-494.
47. Kurachi Y, Ito H, Sugimoto T, Shimizu T, Miki I, Ui M. Pflügers Arch 1989;414:102-104.
48. Needleman P, Turk J, Jakschik BA, Morrison AR, Lefkowith JB. Annu Rev Biochem 1986;55:69-102.
49. Axelrod J, Burch RM, Jelsema CL. Trends Neurosci 1988;11:117-123.
50. Ito H, Takikawa R, Iguchi M, Hamada E, Sugimoto T, Kurachi Y. Pflügers Arch 1990;417:126-128.
51. Yatani A, Okabe K, Polakis P, Halenbeck R, McCormick F, Brown AM. Cell 1990;61:769-776.
52. Heidbüchel H, Callewaert G, Vereecke J, Carmeliet E. Pflügers Arch 1990;416:213-215.
53. Heidbüchel H, Vereecke J, Carmeliet E. PACE 1991;14:648.
54. Kaibara M, Nakajima T, Irisawa H, Giles W. J Physiol Lond 1991;433:589-613.

44. [illegible] 274.

45. [illegible]

46. [illegible]

47. [illegible]

48. [illegible]

49. Axelrod J, [illegible] 123.

50. [illegible]

51. [illegible]

52. [illegible]

53. [illegible]

54. [illegible]

Multiple Cyclic GMP Binding Proteins Involved in the Regulation of Cardiac Calcium Channels

R. Fischmeister, P.-F. Mery, S.M. Lohmann* and U. Walter*

Laboratoire de Physiologie Cellulaire Cardiaque, INSERM U-241, Université de Paris-Sud, F-91405 Orsay, FRANCE

**Medizinische Universitätsklinik, Labor für Klinische Biochemie, Josef Schneider Str. 2, D-8700 Würzburg, FEDERAL REPUBLIC OF GERMANY*

Introduction

There is ample evidence that cAMP and cGMP lead to opposite contractile and electrical responses in the heart (1-10). Since regulation of the L-type calcium channel current (I_{Ca}) often correlates with the regulation of cardiac contraction, it was conceivable that both nucleotides exert opposite actions on I_{Ca}. The stimulatory effect of cAMP on I_{Ca} has been well documented. Cyclic AMP activation of cAMP-dependent protein kinase (cAMP-PK) leads to phosphorylation of L-type Ca^{2+} channels (or a closely associated protein), producing an increase in the mean probability of channel opening and stimulation of macroscopic I_{Ca} (10-12). By symmetry, it has been proposed that the negative inotropic effect of cGMP was mediated by a decrease in I_{Ca}. Among the evidence supporting this hypothesis are the findings that cGMP reduces ^{45}Ca flux (3), shortens action potential duration (13), and inhibits Ca^{2+}-dependent action potentials (14-17). Although initial voltage-clamp studies on multicellular cardiac preparations showed no effect of a photoactivatable cGMP analogue on I_{Ca} (18,19), experiments on isolated cells from frog (20,21), guinea-pig (22,23), rat (24) and embryonic chicken (25) ventricular tissue have provided ample support for an inhibition of I_{Ca} by cGMP. Combination of patch-clamp and intracellular perfusion techniques (20) permitted examination of the effects of internal application of cAMP, cGMP, and their hydrolysis-resistant analogs, as well as of cGMP-dependent protein kinase (cGMP-PK) in isolated cardiac myocytes. Such experiments helped to elucidate the molecular mechanisms involved in cGMP regulation of I_{Ca}. Three different cGMP binding proteins were found to regulate I_{Ca}, with significant species-dependent variations in the extent to which each mechanism contributes to the overall effect of cGMP.

cGMP-Stimulated cAMP-Phosphodiesterase

Cyclic GMP-stimulation of cAMP hydrolysis has been proposed to participate in the cardiac negative inotropic effect of cGMP because (i) cAMP levels were often found to be reduced when cGMP level increased (8,26,27), and (ii) evidence exists for the presence of a cGMP-stimulated cAMP-phosphodiesterase (cGs-PDE) in cardiac tissue (28,29) as well as in myocytes (30). In isolated frog myocytes, the use of poorly hydrolyzable (8Br) analogs in whole-cell patch-clamp studies indicated that cGMP decreased I_{Ca} by stimulation of cGs-PDE because (i) cGMP was without effect on I_{Ca} that had been elevated by 8Br-CAMP, (ii) 8Br-cGMP (which is a much better stimulator of cGMP-PK than cGMP, and a much worse stimulator of cGs-PDE than cGMP) had no effect on I_{Ca}, and (iii) the inhibitory effect of cGMP on cAMP-stimulated I_{Ca} was largely antagonized by isobutyl methylxanthine (IBMX), a nonspecific phosphodiesterase inhibitor (20,21). The hypothesis that I_{Ca} in frog is regulated by cGs-PDE is supported by considerable biochemical and electrophysiological evidence (10). This evidence includes (i) the characterization in frog ventricular cells of a cGs-PDE activity with the same sensitivity to cGMP as I_{Ca} (31) and (ii) a similar inhibition of the effects of cGMP on I_{Ca} and cGs-PDE by high concentrations of IBMX (20,21,31,32).

It is worth mentioning that, in addition to regulating I_{Ca} in frog cardiocytes, cGs-PDE has been recently reported to participate in the response of various non-cardiac cell types to the atrial natriuretic peptide (ANP) (33-35).

cGMP-Dependent Protein Kinase

In contrast to the mechanism of cGMP action in frog heart, the inhibitory effect of cGMP on slow action potentials in guinea pig heart appeared to be independent of changes in cAMP levels (36). Further studies in isolated guinea pig and rat myocytes supported this conclusion, and demonstrated mediation of the cGMP inhibition of I_{Ca} by cGMP-PK (22,24). In order to conclusively demonstrate an effect of cGMP-PK on I_{Ca}, it was necessary to perfuse myocytes with a constitutively active cGMP-PK which did not require cGMP for stimulation so that other effects of cGMP could be avoided. This was achieved by partial trypsinization which removes the first 77 amino acids of cGMP-PK containing the holoenzyme dimerization domain but also an inhibitory domain responsible for inhibition of enzyme activity in the absence of cGMP (37). This treatment reduced the 150 kDa holoenzyme dimer to a monomer of 65 kDa. Intracellular perfusion of 10 - 100 nM cGMP-PK fragment in rat ventricular cells caused 30 - 70% reversal of the I_{Ca} elevated by cAMP or 8Br-cAMP, similar to the extent of inhibition achieved by 0.1 - 1 uM cGMP or 8Br-cGMP (24). The higher concentration of cGMP or analog required in comparison to cGMP-PK may result in part from some degradation of the cyclic nucleotides and from the need for 4 cGMP molecules to fully activate the endogenous cGMP-PK holoenzyme. Studies with ATP-γ-S suggested that cGMP does not activate a phosphatase or inhibit cAMP-PK to reverse the cAMP effects. The cGMP-PK has been shown to stoichiometrically phosphorylate the α_1 and ß subunits of the skeletal muscle Ca^{2+} channel in vitro (38), although these experiments must be extended to the cardiac Ca^{2+} channel in the future. Some other in vitro substrates for cGMP-PK, which have been described, include a 70 kDa protein in rat heart cytosol (39), a 50 kDa protein in guinea pig sarcolemma (40), phospholamban (41) and the ryanodine receptor (42), as well as the inhibitory subunit of troponin (43).

Besides inhibition of I_{Ca} in mammalian cardiocytes, cGMP-PK has been reported to inhibit an amiloride sensitive Na^+ channel in renal inner medullary collecting duct cells (44) and to activate a Ca^{2+} channel in snail neurons (45).

cGMP-Inhibited cAMP-Phosphodiesterase

In a recent study of the effects of several specific inhibitors of the low-K_m cAMP-phosphodiesterases on frog ventricular cell I_{Ca} (32), evidence for the presence of a cGMP-inhibited cAMP-phosphodiesterase (cGi-PDE) in these cells was obtained. Two specific inhibitors (milrinone and indolidan [LY195115]) of this family of enzymes consistently increased I_{Ca} when cAMP was present inside the cells at submaximal concentrations, but not in the absence of cAMP (32). These results suggested a potential role for the cGi-PDE in regulating the amplitude of I_{Ca} upon activation of cGMP production. However, in frog ventricular cells, intracellular application of cGMP only reduced I_{Ca}, suggesting that activation of the cGs-PDE overcomes the inhibition of the cGi-PDE (32). The situation may be different in guinea-pig myocytes, where cGMP-induced inhibition of I_{Ca} occurs through the cGMP-PK and not through the cGs-PDE (22). Indeed, Ono and Trautwein recently found a stimulatory effect of low concentrations of cGMP on cAMP-elevated I_{Ca} in this cell type, while high concentrations of cGMP, 8Br-cGMP or cGMP-PK had either no effect or induced a reduction of I_{Ca} (23). Consistent with the participation of a cGi-PDE in this new effect of cGMP on I_{Ca} are (i) the biochemical characterization of a cGi-PDE in guinea-pig heart (46,47) and (ii) the finding that cGMP exerted no further stimulatory effect on I_{Ca} in the presence of milrinone (23).

Direct Effect of cGMP on Ca^{2+} Channels

cGMP can directly activate a channel in retinal rod photoreceptors (48) and olfactory cells (49), or inhibit a channel in renal inner medullary collecting duct cells (44). Because cGMP was found recently to directly activate the cardiac pacemaker channel, although with a much lower potency than cAMP (50), one may question whether the observed regulation of I_{Ca} by cGMP is due to a direct effect on cardiac Ca^{2+} channels. There is, however, no evidence for the presence of a nucleotide binding site in the reported molecular structure of the cardiac Ca^{2+} channel (51). A direct effect of cGMP on Ca^{2+} channels was unlikely in frog cells because

(i) cGMP inhibited I_{Ca} stimulated by cAMP but not I_{Ca} stimulated by 8Br-cAMP, and (ii) IBMX reversed the inhibitory effect of cGMP (20,21). A similar conclusion, although based on different results, was drawn from our experiments in rat ventricular cells. In this preparation, the cGMP inhibition of I_{Ca} was irreversible when non-hydrolyzable ATP-γ-S was used, suggesting that a phoshorylation-dependent mechanism rather than an direct effect of cGMP on Ca^{2+} channels was involved (24).

Endogenous cGMP Effectors in Myocytes

Evidence exists for the presence of phosphodiesterases and cGMP-PK in myocytes for mediation of the cGMP effects described above. Several different types of phosphodiesterases have been reported in whole heart, and in purified rat myocytes as well, with the exception of the Ca^{2+}-calmodulin activated phosphodiesterase which appeared to derive from some other cell type in the heart (30). Particularly relevant was the presence of both the cGs-PDE and cGi-PDE in rat cardiocytes. The characteristics of different classes of PDEs and their inhibitors (some used clinically) have been recently extensively reviewed (52,53).

Western blot analysis detected low amounts of cGMP-PK in the same isolated rat myocytes which had been used in the patch-clamp analysis of I_{Ca} (24). The calculated concentration of cGMP-PK in myocytes was about 10 fold less than that reported in bovine tracheal smooth muscle (54), and far less than that estimated for other cell types (55,56). If one assumes that the cGMP-PK concentration in vascular smooth muscle of heart may be similar to that in tracheal muscle, and considering the much lower levels of cGMP-PK in cardiac myocytes, it is evident that this has contributed to past difficulties in detecting cGMP-PK in myocytes by immunocytochemistry and in studying cGMP-regulation in the whole heart (8).

cGMP - A Second Messenger in the Heart

The second messenger role of cAMP in cardiac cells is unquestionable. The cAMP-dependent stimulation of I_{Ca} explains most of the positive inotropic effect of several neuromediators and hormones, such as ß-adrenergic agonists (57-59), histamine (60,61), and glucagon (62), as well as agents known to stimulate cAMP synthesis (e.g forskolin [63,64]) or inhibit cAMP degradation (phosphodiesterase inhibitors [31,32,65]).

Although it was in the heart that cGMP levels were first discovered to be physiologically regulated (66), the extent to which this nucleotide participates in the negative inotropic effect of acetylcholine (ACh) (10), as well as other extracellular signals, known to elevate cGMP level in the heart, such as adenosine (67) and ANP (68-71), is unclear (3,4, reviewed in 10,21). There is a similarity, though, in the way cGMP and these agents regulate I_{Ca}. Indeed, in all species but chicken (25), cGMP does not affect basal I_{Ca} but inhibits only I_{Ca} previously stimulated by cAMP-dependent phosphorylation (20-24). The lack of effect on basal I_{Ca} was also characteristic of the inhibitory action of ACh (10,58,72), adenosine (73) and ANP (74,75).

A clear correlation, however, does not exist between the contractile effect of ACh and cGMP levels (10,21). One plausible reason for this is that earlier experiments were carried out on multicellular cardiac preparations which may possess various amounts of cell types other than myocytes, such as endothelial cells and vascular smooth muscle cells, which respond through different mechanisms to ACh and influence each other in a complex manner (76). In addition, these experiments were performed under non voltage-clamp conditions in different cardiac tissues in which ACh activation of the inwardly rectifying K^+ current (77), a source of negative chronotropism and inotropism clearly not related to cGMP (78), would interfere with direct effects of ACh for the regulation of Ca^{2+} channel activity. Conclusive evidence about the participation of cGMP in the effect of various agonists has also been hindered by the presence of a concomitant direct and strong inhibitory effect of these agonists on cardiac adenylyl cyclase (68,69,79-81).

Further experiments at the single cell level are necessary to examine the physiological role of cGMP in the regulation of cardiac cell function. Consequently, we are currently examining the effects on isolated myocytes of various cardiac inotropic agents in the presence or absence of synthetic compounds or cGMP analogs that can specifically distinguish between the various cGMP binding proteins or which can regulate cGMP metabolism.

Acknowledgements

This work was supported by the Deutsche Forschungsgemeinschaft (Ko 210/11-1) and the French/German PROCOPE exchange program.

References

1. Trautwein W, Trube, G. Pflügers Arch 1976;366:293-295.
2. Wilkerson, RD, Paddock, RJ George, WJ. Eur J Pharmacol 1976;36:247-251.
3. Nawrath H. Nature 1977;267:72-74.
4. Diamond J, Ten Eick RE, Trapani AJ. Biochem Biophys Res Commun 1977;79:912-917.
5. Tuganowski W, Kopec P. Naunyn-Schmied Arch Pharmacol 1978;304:211-213.
6. Singh J, Flitney FW. Biochem Pharmacol 1981;30:1475-1481.
7. Goldberg ND, Haddox MK. Ann Rev Biochem 1977;46:823-896.
8. Walter U. Adv Cycl Nucl Prot Phosph Res 1984;17:249-258.
9. Walter U. Rev Physiol Biochem Pharmacol 1989;113:42-88.
10. Hartzell HC. Prog Biophys Mol Biol 1988;52:165-247.
11. Hofmann F, Nastainczyk W, Röhrkasten A, Schneider T, Sieber M. TIPS 1987;8:393-398.
12. Tsien RW, Bean BP, Hess P, Lansman JB, Nilius B, Nowycky, M. J Mol Cell Cardiol 1986;18:691-710.
13. Trautwein W, Taniguchi J, Noma A. Pflügers Arch 1982;392:307-314.
14. Kohlhardt M, Haap K. J Mol Cell Cardiol 1978;10:573-586.
15. Wahler GM, Sperelakis N. J Cyclic Nucleot Prot Phosphoryl Res 1985;10:83-95.
16. Bkaily G, Sperelakis N. Am J Physiol 1985;248:H745-H749.
17. Mehegan JP, Muir WW, Unverferth DV, Fertel RH, McGuirk SM. J Cardiovasc Pharmacol 1985;7:30-35.
18. Nargeot J, Nerbonne JM, Engels J, Lester HA. Proc Natl Acad Sci USA 1983;80:2395-2399.

19. Richard S, Nerbonne JM, Nargeot J, Lester HA. Pflügers Arch 1985;403:312-317.
20. Hartzell HC, Fischmeister R. Nature 1986;323:273-275.
21. Fischmeister R, Hartzell HC. J Physiol (Lond) 1987;387:453-472.
22. Levi RC, Alloatti G, Fischmeister R. Pflügers Arch 1989;413:685-687.
23. Ono K, Trautwein W. J Physiol (Lond) 1992; (in press).
24. Méry P-F, Lohmann SM, Walter U, Fischmeister R. Proc Nat Acad Sci USA 1991;88:1197-1201.
25. Wahler GM, Rusch NJ, Sperelakis N. Can J Physiol 1990;68:531-534.
26. Endoh M. Jap J Pharmacol 1979;29:855-864.
27. Flitney FW, Singh J. J Mol Cell Cardiol 1981;13:963-979.
28. Beavo JA, Hardmann JG, Sutherland EW. J Biol Chem 1971;246:3841-3846.
29. Martins TJ, Mumby MC, Beavo JA. J Biol Chem 1982;257:1973-1979.
30. Bode DC, Kanter JR, Brunton LL. Circ Res 1990;68:1070-1079.
31. Simmons MA, Hartzell HC. Mol Pharmacol 1988;33:664-671.
32. Fischmeister R, Hartzell HC. Mol Pharmacol 1990;38:426-433.
33. Lee MA, West RE, Moss J. J Clin Invest 1988;82:388-393.
34. MacFarland RT, Zelus BD, Beavo JA. J Biol Chem 1991;266:136-142.
35. Whalin ME, Scammel JG, Strada SJ, Thompson W. J Mol Pharmacol 1991;39:711-717.
36. Thakkar J, Tang S-B, Sperelakis N, Wahler GM. Can J Physiol Pharmacol 1988;66:1092-1095.
37. Heil WG, Landgraf W, Hofmann F. Eur J Biochem 1987;168:117-121.
38. Jahn H, Nastainczyk W, Röhrkasten A, Schneider T, Hofmann F. Eur J Biochem 1988;178:535-542.
39. Wrenn RW, Kuo JF Biochem Biophys Res Commun 1981;101:1274-1280.
40. Cuppoletti J, Thakkar J, Sperelakis N, Wahler G. Membr Biochem 1989;7:135-142.
41. Raeymaekers L, Hofmann F, Casteels R. Biochem J 1988;252:269-273.
42. Takasago T, Imagawa T, Furukawa K, Ogurusu T, Shigekawa M. J Biochem 1991;109:163-170.

43. Pfitzer G, Rüegg JC, Flockerzi V, Hofmann F. FEBS Lett 1982;149:171-175.
44. Light DB, Corbin JD, Stanton BA. Nature 1990;344:336-339.
45. Paupardin-Tritsch D, Hammond C, Gerschenfeld HM, Nairn AC, Greengard P. Nature 1986;323:812-814.
46. Reeves ML, Leigh BK, England PJ. Biochem J 1987;241:535-541.
47. Weishaar RE, Kobylarz-Singer DC, Steffen RP, Kaplan HR. Circ Res 1987;61:539-547.
48. Kaupp UB. TINS 1991;14:150-157.
49. Nakamura T, Gold GH. Nature 1987;325:442-444.
50. DiFrancesco D, Tortora P. Nature 1991;351:145-147.
51. Mikami A, Imoto K, Tanabe T, Niidome T, Mori Y, Takeshima H, Narumiya S, Numa S. Nature 1989;340:230-233.
52. Beavo JA, Reifsnyder DH. TIPS 1990;11:150-155.
53. Nicholson CD, Challis RAJ, Shahid M. TIPS 1991;12:19-27.
54. Felbel J, Trockur B, Ecker T, Landgraf W, Hofmann F. J Biol Chem 1988;263:16764-16771.
55. Eigenthaler M, Friedrich C, Schanzenbächer P, Walter U. Eur J Clin Invest 1990;20:A16.
56. Lohmann SM, Walter U, Miller PE, Greengard P, De Camilli P. Proc Natl Acad Sci USA 1981;78:653-657.
57. Kameyama M, Hofmann F, Trautwein W. Pflügers Arch 1985;405:285-293.
58. Fischmeister R, Hartzell HC. J Physiol (Lond) 1986;376:183-202.
59. Hartzell HC, Méry P-F, Fischmeister R, Szabo G. Nature 1991;351:573-576.
60. Hescheler J, Tang M, Jastorff B, Trautwein W. Pflügers Arch 1987;410:23-29.
61. Levi RC, Alloatti G. J Pharmacol Exp Therap 1988;246:377-383.
62. Méry P-F, Brechler V, Pavoine C, Pecker F, Fischmeister R. Nature 1990;345:158-161.
63. Kameyama M, Hescheler J, Hofmann F, Trautwein W. Pflügers Arch 1986;407:123-128.
64. Hartzell HC, Fischmeister R. Mol Pharmacol 1987;32:639-645.
65. Fischmeister R, Hartzell HC. Life Sci 1991;48:2365-2376.

66. George WJ, Polson JB, O'Toole AG, Goldberg N. Proc Natl Acad Sci USA 1970;66:398-403.
67. Singh J, Flitney FW. J Mol Cell Cardiol 1980;12:285-297.
68. Cramb G, Banks R, Rugg EL, Aiton JF. Biochem Biophys Res Commun 1987;148:962-970.
69. McCall D, Fried TA. J Mol Cell Cardiol 1990;22:201-212.
70. Neyses L, Vetter H. Biochem Biophys Res Commun 1989;163:1435-1443.
71. Rugg EL, Aiton JF, Cramb G. Biochem Biophys Res Commun 1989;162:1339-1345.
72. Hescheler J, Kameyama M, Trautwein W. Pfl ügers Arch 1986;407:182-189.
73. Isenberg G, Cerbai E, Klöckner UH. In: Topics and Perspectives in Adenosine Research (eds.) E Gerlach, BF Becker. Springer-Verlag, Berlin Heidelberg, 1987; pp. 323-335.
74. Gisbert M-P, Fischmeister R. Circ Res 1988;62:660-667.
75. Tei M, Horie M, Makita T, Suzuki H, Hazama A, Okada Y, Kawai C. Biochem Biophys Res Commmun 1990;167:413-418.
76. Lohmann SM, Fischmeister R, Walter U. Basic Res Cardiol 1992; (in press).
77. Szabo G, Otero AS. Ann Rev Physiol 1990;52:293-305.
78. Fleming BP, Giles W, Lederer J. J Physiol (Lond) 1981;314-47-64.
79. Belardinelli L, Vogel S, Linden J, Berne RM. J Mol Cell Cardiol 1982;14:291-294.
80. Fischmeister R, Méry P-F, Shrier A, Pavoine C, Brechler V, Pecker F. In: Subcellular Basis of Contractile Failure (eds.) B. Korecky, NS Dhalla. Klüwer Academic Press, Boston, Dordrecht, London, 1990; pp. 39-54.
81. Anand-Srivastava MB, Cantin M. Biochem Biophys Res Commun 1986;138:427-436.

Participation of the Na^+-H^+ Exchange Pathway in Cardiac Pathology

G.N. Pierce and H. Meng

Division of Cardiovascular Sciences, St. Boniface Hospital Research Centre and Department of Physiology, University of Manitoba, Winnipeg, CANADA.

Introduction: Sodium and Cardiac Pathophysiology

Since the pioneering work of Jennings and colleagues (1), the involvement of calcium in myocardial damage and necrosis during pathological conditions is firmly established. However, the role of other ions besides calcium in cardiac pathology is not as well recognized today. Increasing evidence supports a role for sodium in cardiac pathologies as diverse as ischemia/reperfusion (2-6), hypoxia/reoxygenation (7,8), glycosidic toxification (9), the calcium paradox (10,11) and various cardiomyopathies (12-14). In the majority of these disease states, the aberrant sodium homeostasis in the myocardium is not a direct cause of cellular necrosis. Calcium remains as a central necrotic factor in cardiac disease. However, this does not lessen the significance of the change in sodium regulation because this alteration in sodium homeostasis appears to be crucial in triggering the intracellular calcium overload which ultimately leads to cell death. Indeed, it has been suggested that normalization of myocardial sodium levels in the heart during specific noxious challenges can protect the heart by preventing alterations in calcium homeostasis (2-6). Thus, sodium too, appears to have a crucial role in myocardial pathophysiology.

Sodium can cross the myocardial sarcolemmal membrane by several different transport pathways. The potential-dependent Na^+ channel,

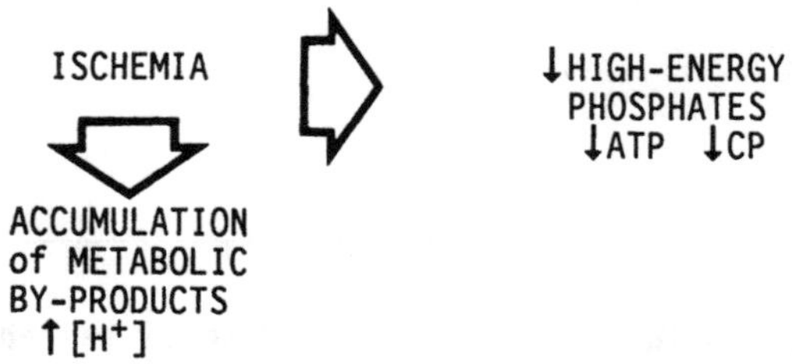

Figure 1. The metabolic changes associated with myocardial ischemia.

the Na^+-H^+ exchanger and the Na^+ pump (Na^+, K^+-ATPase) are the most important transport systems. Other pathways for transarcolemmal sodium movements have been identified (15) but may be of lesser quantitative importance. Na^+ can be compartmentalized within the myocardium (16) (primarily within mitochondria (16) and will be buffered passively by anionic intracellular sites (17)). The Na^+-H^+ exchange system can bring more sodium into the myocardium than any of the other pathways (18). In coupling the Na^+ influx with H^+ efflux, it also functions as an important regulator of myocardial pH. It is stimulated primarily when the intracellular environment is acidic (18). Thus, it is an especially important, active ion transport system during ischemic/reperfusion challenge to the heart when the cellular pH drops considerably (19). Recent data have suggested that activation of the Na^+-H^+ exchanger during ischemic/reperfusion injury may paradoxically lead to cardiac injury rather than maintain heart function and integrity.

Ischemic/Reperfusion Injury in the Heart and the Role of Na^+-H^+ Exchange

Global ischemia results in a dramatic imbalance in the substrate supply/demand equilibrium. As a consequence, two major metabolic changes occur in the heart (Figure 1). First, the cessation of flow eliminates the removal of metabolic end-products like lactate, phosphate and H^+. This leaves the intra- and extra-cellular space about 1 pH unit lower than normal (19). The inadequate substrate supply coupled with continuing energy demands eventually depletes the high energy phosphate stores (1). Secondly, a sequence of ionic events has been hypothesized (2-6,18) to occur during reperfusion of the heart which are initiated by the ischemic acidity. This cascade of reactions is described as:

ischemia

⇩

acidification of the intra and extracellular space

⇩

reperfusion

⇩

rapid washout of the acidic extracellular pH
with a neutral buffered perfusate

⇩

establishment of an outwardly directed H^+ gradient

⇩

stimulation of Na^+-H^+ exchange

⇩

⇧ Na^+ entry into the myocardial cell

⇩

myocardial [Na^+] rises

⇩

stimulation of Na^+-Ca^{2+} exchange

⇩

⇧ Ca^{2+} entry

⇩

intracellular Ca^{2+} overload

⇩

cardiac contractile dysfunction, damage and necrosis

If the cascade of events is correct, then administration of a pharmacological agent which blocks Na^+-H^+ exchange could protect the heart and restore cardiac function and viability. Dimethylamiloride (DMA) is a selective inhibitor of Na^+-H^+ exchange in the heart (2-4). Including 20 uM DMA in the coronary perfusate of the isolated right ventricular wall for 3 minutes prior to 60 minutes of global ischemia and during the first three minutes of reperfusion significantly protects the heart from contracture

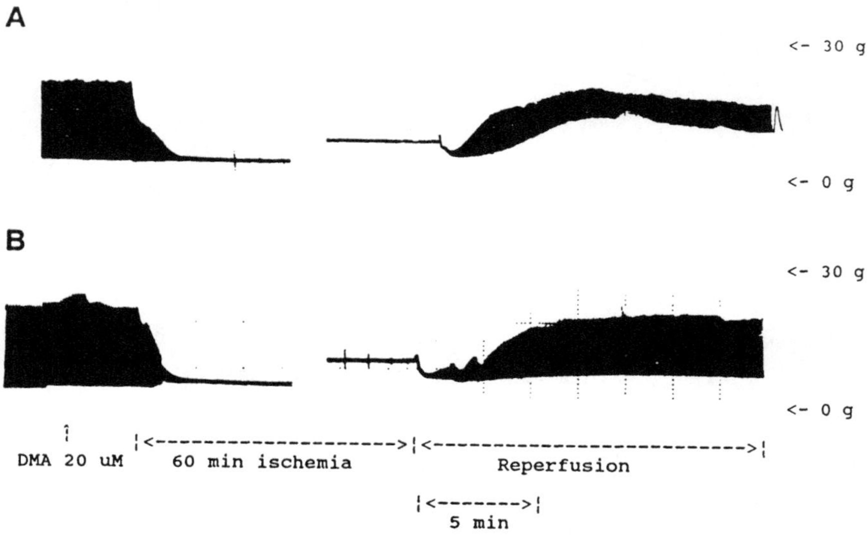

Figure 2. Experimental recordings of the ventricular wall contractile function during ischemia-reperfusion challenge.

Panel A: drug-untreated ventricular wall. Note the resting tension rises and the recovery of developed tension is poor during reperfusion.

Panel B: DMA treatment for 3 min prior to and after ischemia. The rise in resting tension is prevented and the recovery of developed tension is improved by the application of DMA.

formation and allows for a much better recovery of active developed tension (Figure 2). The [DMA] employed (20 uM) is near the IC_{50} of this drug for the inhibition of Na^+-H^+ exchange (~ 7 uM (2)). Inclusion of the drug in the myocardial perfusate also protects the heart from structural damage. Creatine phosphokinase (CPK) is an enzyme normally localized to the intracellular space. Its presence in the coronary effluent reflects damage to the integrity of the cell wall. During reperfusion in the absence of any drug treatment, CPK activity in the coronary effluent rises many fold (Figure 3). DMA treatment normalizes this rise in CPK activity.

These data are consistent with an important role for Na^+-H^+ exchange in ischemic/reperfusion injury in the heart. The action of DMA is sensitive to Na^+ (3) and pH (4) which again suggests that

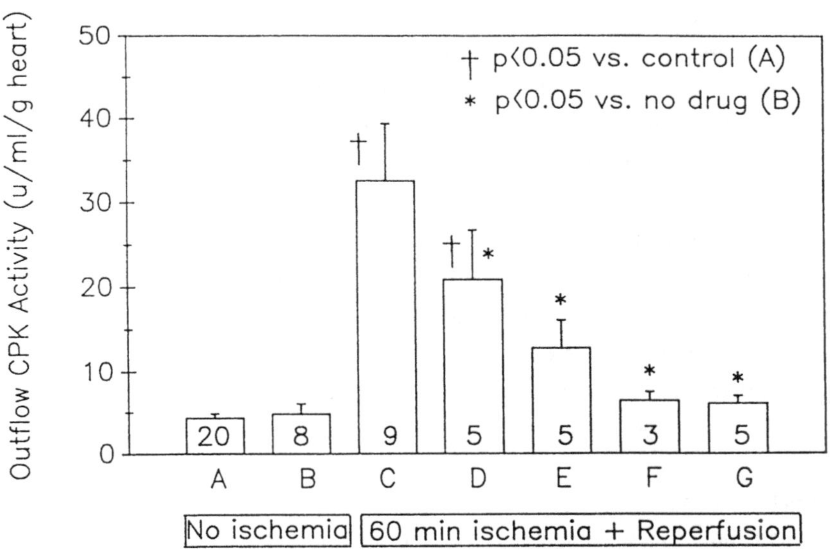

Figure 3. Coronary effluent creatine phosphokinase (CPK) activity in the perfused right ventricular wall. A: control; B: Control + 20uM DMA; C: 6 min reperfusion; D: C+1 uM DMA for 3 minutes; E: C+5 uM DMA for 3 minutes; F: C+20 uM DMA for 3 minutes; G: C+20 uM DMA for 30 minutes. | P<0.05 vs. A; *P<0.05 vs. C.

its effects primarily are via an inhibition of the Na^+-H^+ exchange pathway. For example, administration of a relatively low (35 uM), subtoxic dose of ouabain to block the Na^+ pump and thereby increase intracellular Na^+, worsens the damage to the heart during reperfusion (3). DMA, even in the presence of ouabain, is capable of negating the effects of ouabain and normalizing the increase in intracellular Na^+ (3). Several other characteristics of DMA's actions argue against an important effect on any other myocardial ion transport pathway (2-4).

Hypoxic/Reoxygenation Injury in the Heart

The data presented here exhibit some similarities to those found in another experimental model of pathology: hypoxia/reoxygenation. For example, administration of a low, subtoxic dose of ouabain to hypoxic hearts results in poorer force development and an elevated end diastolic pressure upon reoxygenation when compared to drug untreated preparations (7). Furthermore, the rise in intracellular Na^{+} and Ca^{2+} content in hypoxic hearts can be blocked by administration of ethylisopropylamiloride (EIPA) (8). EIPA is a potent inhibitor of Na^{+}-H^{+} exchange (8). These data suggest that the hypoxic acidification of the myocardium may induce ionic alterations similar to those projected in the present work for ischemic/reperfusion challenge to the heart and may indicate a role for Na^{+}-H^{+} exchange in hypoxic/reoxygenation injury. Further study of other disease states which may exhibit alterations in intracellular pH is warranted to determine the extent of involvement of Na^{+}-H^{+} exchange in the pathology.

Acknowledgements

This work was supported by the Heart and Stroke Foundation of Manitoba. G.N. Pierce was a Basic Science Career Investigator of the Manitoba Medical Service Foundation.

References

1. Jennings RB, Reimer KA. Am J Pathol 1981;102:241-245.
2. Meng H, Pierce GN. Am J Physiol 1990;27:H1615-H1619.
3. Meng H, Pierce GN. J Pharmacol Exp Ther 1991;256:1094-1100.
4. Meng H, Lonsberry BB, Pierce GN. J Pharmacol Exp Ther 1991;256: (in press).
5. Karmazyn M. Am J Physiol 1988;255:H608-H615.
6. Tani M, Neely JR. Circ Res 1989;65:1045-1056.
7. Cunningham MJ, Apsttrin CS, Weinberg EO, Lorell BH. Am J Physiol 1989;25:H681-H687.
8. Anderson SE, Murphy E, Steenbergen C, London RE, Cala PM. Am J Physiol 1990;259:C940-C948.
9. Waldorff S, Hansen PB, Kjaergard H, Buch J, Egeblad H, Steiness E. Clin Pharmacol Ther 1981;30:172-176.
10. Pierce GN, Maddaford TG, Kroeger EA, Cragoe EJ. Am J Physiol 1990;258:H17-H23.
11. Alto LE, Dhalla NS. Am J Physiol. 1979;237:H713-H719.

12. Pierce GN, Ramjiawan B, Dhalla NS, Ferrari R. Am J Physiol 1990;258:H255-H261.
13. Makino N, Jasmin G, Beamish RE, Dhalla NS. Biochem Biophys Res Commun 1985;133:491-497.
14. Makino N, Dhuvarajan R, Elimban V, Beamish RE, Dhalla NS. Can J Cardiol 1985;1:225-232.
15. Wright EM. Ann Rev Physiol 1985;47:127-141.
16. Vaghy PL, Johnson JD, Matlib MA, Wang T, Schwartz A. J Biol Chem 1982;257:6000-6002.
17. Reeves JP, Sutko JL. J Biol Chem 1983;258:3178-3182.
18. Lazdunski M, Frelin C, Vigne P. J Mol Cell Cardiol 1985;17:1029-1042.
19. Poole-Wilson PA. J Mol Cell Cardiol 1978;10:511-526.

Phospholipase D: A New Avenue to the Phospholipid Signalling Pathways in the Myocardium

J.T.A. Meij and V. Panagia

Division of Cardiovascular Sciences, St. Boniface General Hospital Research Centre and Department of Anatomy, University of Manitoba, Winnipeg, Manitoba, CANADA

Over the past decade it has become known that phospholipids, present in the plasma membrane, play an essential part in receptor signal transduction. Firstly, changes in plasma membrane phospholipid composition alter the physicochemical properties of the lipid bilayer and the lipid domains of membrane proteins (receptors, ion channels, enzymes etc.), thereby influencing cell function [1]. Secondly, phospholipids originate several intracellular messenger molecules [2-5]. These phospholipid metabolites may stem from the headgroup, such as inositol 1,4,5-trisphosphate (Ins(1,4,5)P_3) and inositol 1,3,4,5-tetrakisphosphate (Ins(1,3,4,5)P_4), or they may be of a lipid nature, such as certain non-esterified fatty acids (NEFAs), lysophospholipids (lysoPLs), sn-1,2-diacylglycerol (DAG), phosphatidic acid (PtdOH) and eicosanoids. The presence of the enzymes responsible for the formation of above mentioned metabolites in myocardial cells has in most cases been confirmed only recently [6-9]. Moreover, with regard to the knowledge on the mechanisms of stimulus-response coupling through phospholipid pathways, only a corner of the veil has been lifted.

NEFAs and lysoPLs can be generated directly by the actions of phospholipases A (PL A). A typical case is the PL A_2-catalyzed liberation of arachidonic acid, both a messenger itself and the precursor for eicosanoids [3]. Although there are some indications that PL A_2 may be activated via receptors (bradykinin, angiotensin II, vasopressin and ß-adrenergic) [10], the involvement of PL A in signal transduction is still doubtful.

Currently, the most intensively studied signalling pathway is the phosphoinositide (PI) cycle. Several receptors, among which are the myocardial muscarinic [11,12] and α_1 -adrenergic [13,14] receptors, are coupled to a sarcolemmal (SL) phospholipase C (PL C) that specifically hydrolyzes phosphatidylinositol 4,5-bisphosphate (PtdIns(4,5)P_2) to yield Ins(1,4,5)P_3 and DAG. Various actions in myocardial cells of these two second messengers as well as their further processing, have been described [for reviews see 15-17]. Briefly, it was shown that Ins(1,4,5)P_3 alone or in concert with Ins(1,3,4,5)P_4, is able to (enhance the) release of Ca^{2+} ions from intracellular stores in the sarcoplasmic reticulum (SR) [18], whereas DAG is one of the factors necessary for the activation of protein kinase C (PKC) [19].

Recently, an alternative pathway for the generation of DAG has been evident in the subsequential activities of a phospholipase D (PL D) which cleaves phospholipids to the headgroup alcohol and PtdOH and of a phosphatidate phosphohydrolase (PPH) which dephosphorylates PtdOH to DAG [20]. The present paper will focus on the latter pathway. In view of the current abundance of excellent reviews on PL D and its role in phosphatidylcholine (PtdCho) metabolism [21-26], we will give only a brief overview of general findings in mammalian tissues and go into more detail on findings in the myocardium.

General Characteristics of the Mammalian Phospholipase D

PL D is an enzyme that modifies the headgroup of phospholipids. It may transfer a hydroxyl group to the phosphatidyl moiety (hydrolysis) thereby forming PtdOH, or it may transfer a short chain alcohol (transphosphatidylation), e.g. ethanol or glycerol, producing Ptd-ethanol and Ptd-glycerol, respectively [20]. Because the capacity to catalyze transphosphatidylation is unique to PL D, it is often used to distinguish PL D from other hydrolytic activities [20].

PL D activities have been identified in plants, animals and

microorganisms. In mammalian tissues, its activity was first shown in rat brain by Saito and Kanfer in 1973 [27], and subsequently extensively studied in that tissue. The main findings of these studies were that mammalian PL D was a membrane-bound enzyme which could be active towards phosphatidylinositol (PtdIns) or phosphatidylethanolamine (PtdEtn), but foremost preferred PtdCho [28]. In all other tissues and organs examined by Kanfer and associates, PL D was present and displayed similar biochemical characteristics [29]. Since then, PL D activities have been reported in various tissues with different biochemical properties [5]. Although it seems that multiple isozymes of PL D exist, their role in cellular signalling remains to be elucidated.

Phospholipase D in the Myocardium

The first detection of PL D activity in the heart was done by Kanfer and collaborators when surveying microsomal preparations derived from a variety of rat tissues [29]. The heart PL D was demonstrated with exogenous substrate both in the hydrolytic and the transphosphatidylation modes [29-30]. Further evidence for the presence of this enzyme, and indications for its involvement in stimulus-response coupling, came from studies in perfused chicken heart done by Löffelholz and coworkers [31-33]. They found that choline efflux in the perfusate of the heart could not be blocked completely by the PL A_2 inhibitor, mepacrine [31,32]. They, furthermore, demonstrated that the muscarinic agonist carbachol increased the mepacrine-insensitive choline-efflux [31,32]. Because both basal and agonist-stimulated choline efflux were Ca^{2+}-independent [32], both PL A_2 and PL C had to be excluded as mechanisms and thus leaving PL D. This supposition was confirmed when the same group of investigators showed concomitant increases in PtdOH and choline after carbachol or oleate perfusion, also under Ca^{2+}-free conditions [33]. Conclusive evidence for the presence of a PL D activity in the myocardium was provided by in vitro studies in our laboratory [9]. Examination of the subcellular distribution of the enzyme showed that a PtdCho-specific, PtdOH-producing enzyme with transphosphatidylation activity was present in several membrane

Table I Characteristics of the myocardial phospholipase D

The enzyme:

- is membrane-bound, present in various subcellular membrane fractions
- is capable of transphosphatidylation
- has a preference for PtdCho as a substrate
- requires unsaturated fatty acids (5 mM sodium oleate) for maximum activation in vitro
- is active in the absence of Ca^{2+}
- has an optimum pH of 6.5
- has an in vitro activity that is linear over 90 min at 25°C.

fractions, but absent in the cytosolic fraction. Determination of PL D activity in purified myocardial subcellular organelles resulted in the following enrichment factors: 14-fold in SL, 11-fold in SR, and 5-fold in mitochondrial membranes [9]. These results and subcellular distribution studies indicated that the surface membrane contains most of the PL D activity, although a considerable portion of the activity is located intracellularly, away from membrane surface receptors. In all these subcellular preparations a massive formation of DAG accompanied PtdOH formation which, in SL, comprised only 6% of the total products [9]. Whereas the transphosphatidylation activity and the PtdOH formation (in absence of ATP) could be ascribed completely to PL D activity, DAG might have been generated along two PtdOH hydrolyzing pathways: directly via PtdCho-specific PL C, or in two subsequential steps by PL D/PPH. Inclusion in the assay medium of potassium fluoride (KF, 25 mM), which is known to partially inhibit PPH, increased the amount of PtdOH optimally to 80% of total PtdCho breakdown products in both SL and SR membranes [9]. This indicates that DAG formed is the product of PPH activity, which was blocked - although incompletely - by KF. Another mode to change the PtdOH/DAG ratio in both SL and SR membranes, was to lower the incubation temperature from 30 °C to 25 °C. Under these conditions PtdOH comprised 90% of the total products [9]. Both factors (KF, lower temperature) decreased the total PtdCho-hydrolysis, but had a more drastic effect in depressing PPH. Table I summarizes the main properties of PL D in rat myocardial SL and SR membranes. PL D activity was also detected in canine, bovine, porcine and guinea pig myocardial SL, expressing species related differences in terms of catalytic rate [unpublished results]. More recent studies on the biochemical characteristics of rat myocardial SL and SR PL D showed a depression induced by hydrogen peroxide [34] and sulfhydryl group dependence of the enzyme [35].

Mode of Activation of Phospholipase D

Despite the extensive work that Kanfer and coworkers did on PL D, it was not until very recently that a possible mechanism for the activation of PL D emerged. More and more reports came out showing that PL D could be activated in response to cell stimulation, either through agonist-receptor binding or with phorbol ester [for a listing see 23]. Furthermore, several studies in which PL D activity was enhanced by guanosine 5'-O-(-3-thiotriphosphate) (GTPγS), suggest that a GTP-binding protein may be involved in the receptor-activation of PL D [36-38].

The key problem in defining the manner of PL D activation, however, is whether PL D is linked directly to a receptor (possibly via a G-protein) or is activated indirectly by PI-PL C products. This question arose from the fact that many of the receptors found to stimulate PL D, had earlier been found to be coupled to PI-PL C [23]. The indirect mechanism is supported by many studies demonstrating a role for PKC and PKC-activators as well as for Ca^{2+} in the stimulation of PL D [5]. Based on this close-knit relationship between PI-PL C and PtdCho-PL D a model has been proposed, in which PL D activation is a consequence of PI-PL C stimulation [5]. As depicted in Figure 1, either DAG-stimulated PKC or Ins(1,4,5)P_3-mobilized Ca^{2+}-ions or the two together, activate PL D. Both the PtdOH and the DAG (via PPH) produced from PtdCho-hydrolysis provide mechanism of positive feedback to maintain PL D activity. DAG does

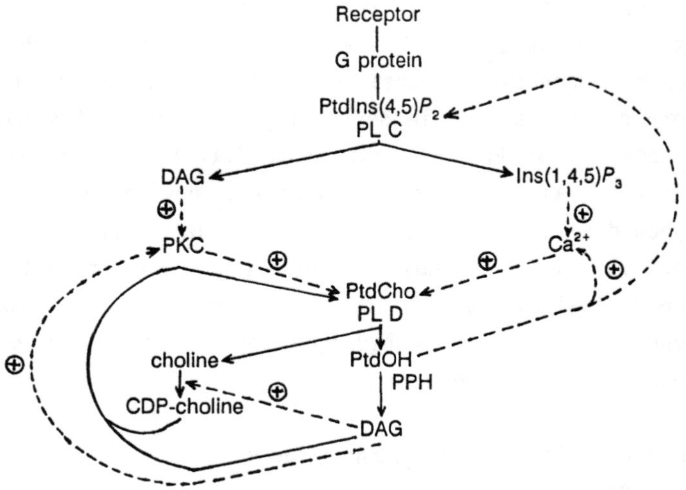

Figure 1. Proposed mechanism for the activation of phospholipase D

this through its effect on PKC, whereas PtdOH does this through its reported effects as a Ca^{2+}-mobilizing agent [5] and/or as activator of PI-PL C [39]. The mechanisms discussed are, however, based on results obtained in non-myocardial cells. Which receptors in the heart are linked to PL D has as yet to be determined. The indications for the muscarinic receptor are strong, as the earlier described experiments of Löffelholz and coworkers showed [31-33]. The evidence for the ß-adrenoceptors is less clear. The same group of investigators performed similar experiments with cAMP-increasing agents (ß-adrenergic agonist, forskolin, 3-isobutyl-1-methylxanthine) but found that, although the choline efflux was elevated above basal level, the increase was completely blocked by either mepacrine or by a low Ca^{2+} concentration in the perfusion medium [40]. Therefore, the ß-adrenoceptor-induced choline efflux seems to be due to cAMP/Ca^{2+}-stimulated PL A_2 activity. Other myocardial receptor types, such as the P_{2y}-purinergic and vasopressin receptors (coupled to PL D in endothelial cells and hepatocytes, respectively [5]), have yet to be tested. A further possibility could be, according to the model illustrated in Figure 1, the myocardial α_1-adrenoceptors. Since these are coupled to PI-PL C [12] they could be indirectly linked to PL D as well.

Apart from indirect stimulation via PL C, a similar mechanism via PL A is equally possible. The finding that in most tissues PL D activity is latent and unmasked by fatty acids [30], points towards this. The optimum oleate concentration for the hydrolysis of both exogenous and endogenous substrate in rat brain microsomes was 5 mM. The same value was also found to be optimum for the in vitro PL D hydrolysis of exogenous substrate in rat heart membranes [9]. These high concentrations seemed to exclude activation of PL D by PL A-induced fatty acid liberation. However, in the perfusion experiments by Löffelholz and coworkers showing oleate induced choline-efflux, only 20 uM oleate was used [33]. Recent findings in our laboratory also show that for the PL D hydrolysis of endogenous PtdCho substrate a much lower concentration of oleate is required [unpublished results]. Furthermore, Kanfer and associates were able to show that the PL A_2 activator melittin was capable of indirectly activating PL D [20]. Therefore, a role for PL A in the activation of PL D cannot be excluded.

Functional Significance of Phospholipase D Activation

Since it is presently unclear under which circumstances PL D is activated, the physiological role of the enzyme cannot fully be defined. Considering the effects that the PtdCho-derived messenger molecules are believed to have, the possibilities are many. Reports on treatment of myocardial preparations with exogenous PL D [41-45] may provide some information, but the described effects could have been due to membrane perturbation and/or PtdCho depletion.

In rat heart, exogenous PL D produced an increase in Ca^{2+} binding to the SL, concomitant with a positive inotropic effect [41]. Also, in rabbit ventricular muscle, PL D produced an increase in contractile force [42]. Furthermore, PL D stimulated Na^+-Ca^{2+} exchange in cardiac SL vesicles [43] and generated Ca^{2+}-dependent slow action potentials in rat atrium [44]. The finding that the latter effect could also be elicited by PtdOH [44] and a further report that PtdOH is able to stimulate the SL Ca^{2+} pump [46] suggest that the production of PtdOH, more than membrane perturbation, is responsible for the observed effects. Effects on intracellular membranes have been reported as well. PtdOH was shown to release Ca^{2+} from SR [47] and PL D treatment caused a reduction of Ca^{2+} uptake in fragmented SR, in parallel to phospholipid cleavage [45]. Altogether these observations support a role for PL D-derived PtdOH in the regulation of Ca^{2+} movements.

PtdOH may also interfere with other Ca^{2+} mobilizing pathways by modulating cyclic nucleotide phosphodiesterase activity [48]. Furthermore, in other cell types several effects of PtdOH have been described [5] which may also occur in cardiac cells. The most remarkable of these are the inhibition of adenylate cyclase [49] and the activation of PI-PL C [39].

Yet, as described earlier, under optimum in vitro conditions and in the absence of phosphatase inhibitor, the main product of myocardial PtdCho hydrolysis is not PtdOH but DAG. At present it is controversial whether or not PtdCho is a quantitatively more important source of DAG than the phosphoinositides [50,51]. The feedback stimulatory cycle via PKC (Fig. 1) would result in a more prolonged DAG formation from PL D/PPH than from PI-PL C, which is subject to a feedback inhibitory mechanism [52]. Thus, PL D catalyzed PtdCho hydrolysis may provide a mechanism to sustain stimulation of PKC and to elicit long-term responses.

It has been shown that several PKC subspecies exist which vary in properties including sensitivity to DAG molecular species [53]. Generally, DAG species with unsaturated fatty acyl groups were shown to be more effective activators of PKC [54]. Since each phospholipid has its own characteristic fatty acyl composition, the DAG molecular species and therefore the PKC isoforms that are activated, may depend on whether the DAG is derived from PtdCho or from PtdIns(4,5)P_2. Previously, it was thought that phosphoinositide-derived DAG molecules were as exceptionally rich in arachidonate as PtdIns [23]. Consequently, phosphoinositide-derived DAG was considered to be a potentially strong PKC activator. Recent studies, however, revealed that PtdIns 4-monophosphate and PtdIns(4,5)P_2, the substrate(s) for PI-PL C, have a much lower unsaturated fatty acid content than PtdIns [55]. Accordingly, the differences in DAG species produced by PtdCho-PL D and by PI-PL C are much more subtle than presumed earlier. Nevertheless, the PKC isoforms activated and the pattern of their activation may still be an element of difference.

The relatively low arachidonic acid content in PtdIns(4,5)P_2 [55] also eliminates the old supposition that phosphoinositide-derived DAG is a rich source for eicosanoid production. PtdCho-derived DAG may serve this purpose equally well. It is also possible that the PtdCho-PL D activity, alone or in association with PPH, contributes to the synthesis of triacylglycerol or phospholipids via PtdOH or DAG intermediates [56]. It has been suggested that DAG causes the translocation from cytosol to SR of cytidylyl transferase, the rate-limiting enzyme in the CDP-choline pathway of PtdCho synthesis [22]. Together with the increased cytosolic amounts of choline, another product of PtdCho-PL D activity, this could promote the resynthesis of PtdCho at the SR [57]. These newly formed PtdCho molecules in the SR membrane would then be available for intermembrane translocation by phospholipid transfer proteins [58,59]. Such a process can be regarded as a means to restore the depleted PtdCho levels of the cardiac membranes, in particular the SL membrane.

Conclusion

As obvious from the above discussion, most of the questions regarding myocardial PL D are still unanswered. Its involvement in signal transduction seems clear, but receptors linked to it have yet to be determined. In addition, to fully understand the physiological role of PtdCho-hydrolyzing PL D it is necessary to establish which is the main in vivo product, PtdOH or DAG, and what is the duration of their presence in the stimulated myocardial cell.

Acknowledgements

The support to Dr. V. Panagia by the Medical Research Council of Canada and the Manitoba Heart and Stroke Foundation is gratefully acknowledged. Dr. J.T.A. Meij was a Fellow of the Medical Research Council of Canada.

References

1. Shinitzky M. In: Physiology of membrane fluidity I (ed.) M Shinitzky. CRC Press, Boca Raton, USA 1984; pp. 1-51.
2. Crews FT. In: Phospholipids and cellular regulation I (ed.) M Kuo. CRC Press, Boca Raton, USA 1985;131-158.
3. Dennis EA. Bio/Technology 1987;5:1294-1300.
4. Rana S, Hokin LE. Phys Rev 1990;70:115-162.
5. Exton JH. J Biol Chem, 1990;265:1-4.
6. Franson RC, Pang DC, Towle DW, Weglicki W. J Mol Cell Cardiol 1978;10:921-930.
7. Panagia V, Ganguly PK, Dhalla NS. Biochim Biophys Acta 1984;792:245-253.
8. Edes I, Kranias EG. Basic Res Cardiol 1989;85:78-87.
9. Panagia V, Ou C, Taira Y, Dai J, Dhalla NS. Biochim Biophys Acta 1991;1064:242-250.
10. Axelrod J. Biochem Soc Trans 1990;18:503-507.
11. Quist EE. Biochem Pharmacol 1982;31:3130-3133.
12. Brown SL, Brown JH. Mol Pharmacol 1983;24:351-356.
13. Gaut ZN, Huggins CG. Nature 1966;212:612-613.
14. Brown JH, Buxton IL, Brunton LL. Circ Res 1985;57:532-537.
15. Brown JH, Jones LG. In: Phosphoinositides and receptor mechanisms (ed.) JW Putney. Alan R. Liss, Inc., New York, USA 1986; pp. 245-270.
16. Volpe P, DiVirgilio F, Bruschi G, Regolisti G, Pozzan T. In: Inositol lipids in cell signalling (eds.) RH Michell, AH Drummond, CP Downes. Academic Press, London, UK 1989; pp. 377-404.
17. Meij JTA, Panagia V. In: Catecholamines and heart disease (ed.) PK Ganguly. CRC Press, Boca Raton, USA 1991; pp. 245-256.

18. Fabiato A. Biophys J 1986;49:190 (Abstr).
19. Nishizuka Y. Nature 1984;308:693-697.
20. Kanfer JN. In: Phosphatidylcholine metabolism (ed.) DE Vance. CRC Press, Boca Raton, USA 1989; pp. 65-86.
21. Löffelholz K. Biochem Pharmacol 1989;38:1543-1549.
22. Pelech SL, Vance DE. Trends Biochem Sci 1989;14:28-30.
23. Billah MM, Anthes JC. Biochem J 1990;269:281-291.
24. Billah MM, Anthes JC, Egan RW, Mullmann TJ, Siegel MI. In: Biology of Cellular Transducing Signals (ed.) J Vanderhoek. Plenum Press, New York, USA 1990; pp. 301-311.
25. Shukla SD, Halenda SP. Life Sci 1991;48:851-866.
26. Dennis EA, Rhee SG, Billah MM, Hannun YA. FASEB J 1991;5:2068-2077.
27. Saito M, Kanfer JN. Biochem Biophys Res Commun 1973;53:391-398.
28. Taki T, Kanfer JN. J Biol Chem 1979;254:9761-9765.
29. Chalifour RJ, Kanfer JN. Biochem Biophys Res Commun 1980;96:742-747.
30. Chalifour RJ, Kanfer JN. J Neurochem 1982;39:299-305.
31. Corradetti R, Lindmar R, Löffelholz K. J Pharmacol Exp Ther 1983;226:826-832.
32. Lindmar R, Löffelholz K, Sandmann J. Naunyn-Schmiedeberg's Arch Pharmacol 1986;332:224-229.
33. Lindmar R, Löffelholz K, Sandmann J. Biochem Pharmacol 1988;37:4689-4695.
34. Dai J, Panagia V. J Mol Cell Cardiol 1991;23(Suppl III):S41 (Abstr).
35. Dai J, Padua R, Panagia V. J Mol Cell Cardiol 1991;23(Suppl V):S98 (Abstr).
36. Irving HR, Exton JH. J Biol Chem 1987;262:3440-3443.
37. Bocckino SB, Blackmore PF, Wilson PB, Exton JH. J Biol Chem 1989;262:15309-15315.
38. Martin TW, Michaelis KC. J Biol Chem 1989;264:8847-8856.
39. Jackowski S, Rock CO. Arch Biochem Biophys 1989;268:516-524.
40. Lindmar R, Löffelholz K, Sandmann J. Naunyn-Schmiedeberg's Arch Pharmacol 1986;334:228-233.

41. Burt JM, Rich TL, Langer GA. Am J Physiol 1984;247:H880-H885.
42. Langer GA, Rich TL. Circ Res 1985;56:146-149.
43. Philipson KD, Nishimoto AY. J Biol Chem 1984;259:16-19.
44. Knabb MT, Rubio R, Berne RM. Pflügers Arch-Eur J Physiol 1984;401:435-437.
45. Fiehn W. Lipids 1978;13:264-266.
46. Carafoli E. In: Calcium Antagonists and Cardiovascular Disease (ed.) L. H. Opie. Raven Press, New York, USA 1984; pp. 29-41.
47. Limas CJ. Biochem Biophys Res Commun 1980;95:541-546.
48. Dubois M, Némoz G, Lagarde M, Prigent A-F. Biochem Biophys Res Commun 1991;170:800-809.
49. Murayama T, Ui M. J Biol Chem 1985;260:7226-7233.
50. Kennerly DA. J Immunol 1990;144:3912-3919.
51. Allan CJ, Exton JH. FASEB J 1991;5:A790 (Abstr).
52. Meij JTA, Lamers JMJ. J Mol Cell Cardiol 1989;21:661-668.
53. Qu Y, Torchia J, Phan TD, Sen AK. Mol Cell Biochem 1991;103:171-180.
54. Go M, Sekiguchi K, Nomura H, Kikkawa U, Nishizuka Y. Biochem Biophys Res Commun 1987;144:598-605.
55. Lamers JMJ, Dekkers DHW, Mesaeli N, Meij JTA, Panagia V, Van Heugten HAA. J Mol Cell Cardiol 1991;23(Suppl V):S100 (Abstr).
56. Tijburg LBM, Geelen MJH, Van Golde LMG. Biochim Biophys Acta 1985;1004:1-9.
57. Hatch GM, O K, Choy PC. Biochem Cell Biol 1989;67:67-77.
58. Kagawa Y, Johnson LW, Racker E. Biochem Biophys Res Commun 1973;50:245-251.
59. Venuti SE, Helmkamp GM. Biochim Biophys Acta 1988;946:119-128.

C. CARDIAC ELECTRIC FIELD

Cardiac Pacemaker Activity: From Single Cells to Modelling the Heart

D. Noble[1,2], J.C. Denyer[1], H.F. Brown[1], R. Winslow[2,3], A. Kimball[4]

[1]University Laboratory of Physiology, Parks Road, Oxford, OX1 3PT, UNITED KINGDOM, [2]Army High Performance Computer Center, University of Minnesota, USA, [3]Department of Biomedical Engineering, The Johns Hopkins University School of Medicine, Baltimore, MD 21205, USA and [4]Thinking Machines Corporation, Cambridge, MA 02142, USA

Introduction

Since the introduction of the methods of cell isolation, patch clamping, cell dialysis and intracellular ion indicators, very major advances have been made in understanding the electrophysiology of cardiac cells (for recent reviews see Noble, 1984; Noble & Powell, 1987; Irisawa, Brown & Giles, 1992). As the information has been made sufficiently precise, it has been incorporated into mathematical models, starting with the Purkinje conducting system (DiFrancesco & Noble, 1985), and extending subsequently to sinus node cells (Noble & Noble, 1984; Noble, DiFrancesco & Denyer, 1989), atrial cells (Hilgemann & Noble, 1987; Earm & Noble, 1990) and ventricular cells Noble et al, 1991). These models have proved successful not only in reproducing normal cardiac electrical activity, but also in reconstructing some of the cellular mechanisms of arrhythmia, including ectopic beating in low potassium and cardiac glycosides (Noble, 1991), action potential shortening during ATP depletion (Nichols & Lederer, 1990), and the early after-depolarizations characteristic of potassium blockers and calcium agonists (in preparation).

The rhythm and arrhythmias of the heart depend not only on these cellular mechanisms but also on multicellular interactions. The heart is a syncytium in which cellular interaction occurs not only with near neighbours, but also across the whole tissue (see for example, Kirchoff et al, 1987; Janse & Wit, 1989). It is therefore important not only to construct models of the individual cells in different regions of the heart but also to incorporate these into massive network models. Until recently, such work was seriously inhibited by the computing power required. For the latest generation of cellular models with up to 30 simultaneous non-linear differential equations to solve, several minutes are required to compute a few seconds of activity even on some of the fasted IBM machines. The computing time required for networks of anything more than a few cells then becomes prohibitive. Yet there are over 100,000 cells in the sinus node, several million in the atrium, and hundreds of millions in the ventricle. Of course, it may well be necessary to model on such a grand scale with the unit of modelling being a single cell. But it will certainly be necessary to model with at least tens of thousands of units, and preferably millions, if the multicellular mechanisms of arrhythmia and fibrillation are to be tackled effectively.

The requisite computing power is provided by the new generation of super-computers, and very particularly by massively parallel computers. These are machines with up to tens of thousands of processors. The machine used in the work reported in this article is the Connection Machine 200 (Thinking Machines Inc), a 32,000 processor computer installed at the US Army High Performance Computing Research Centre at the University of Minnesota. Feasibility studies we have done on this machine show that networks of over 4 million cells could be investigated. The results we describe in this paper use networks of up to 64,000 cells. The first problem we have chosen to tackle is the mammalian sinoatrial node and its interaction with the surrounding atrial tissue. There are two reasons for this choice. The first is that the experimental results obtained from cells in these two areas of the heart permit very accurate modelling to be performed, including the variation of properties known to occur between the different anatomical regions of the node. The second is that the mammalian sinus node, e.g. that of the rabbit, contains at most 200,000 - 400,000 cells (considering the central 0.1 mm^2 has been estimated to contain 5000 cells (Bleeker et al, 1980)). A network of 64,000 cells is therefore potentially realistic and full use of the machine's capacity certainly could be. We will first review the development of work on the single cell modelling and then report on progress being made in incorporating these models into massive networks.

Single Cell Modelling

Since Noma and Irisawa (1976) first succeeded in voltage clamping very small dissected preparations from the mammalian sinoatrial node, a very detailed picture has been developed of the conductance mechanisms involved in the natural pacemaker of the heart (Irisawa, Brown and Giles 1992). The early experiments were very difficult to perform since the successful penetration of such small cells with two microelectrodes was not easy. Moreover, it was impossible to tell what proportion of the cells in the multicellular preparation were reasonably normal and what proportion had suffered from the dissection procedure. Even though it was likely that damaged cells simply disconnected themselves from the remaining intact network (via calcium-induced closure of the nexus channels), so leaving those cells that were connected to the clamp electrodes reasonably normal, this problem made it difficult to extrapolate from the properties of the small network to the properties of individual cells. We will refer later in this article to an important calculation (Noble, 1982) that is particularly dependent on this kind of extrapolation.

It was, therefore, of prime importance when the technique of single isolated cells became available (see Noble and Powell, 1987, for reviews of these techniques and their results) to determine whether the method of enzyme digestion of the connective tissue could be used on the tiniest cardiac cells of all; the sinus node cells. The first results were reported by Tanaguchi, Kokubun, Noma and Irisawa (1981 - for other references to early work see Denyer and

Brown, 1990a, b, c) using cells that seemed to have normal electrophysiological properties but which were, however, abnormal anatomically; their appearance was round rather than spindle shape. Isolated single cells tend to round up when calcium balance is disturbed, so this suggests both that the cells may have been at least partially calcium overloaded and, since their rhythmic properties were not obviously very different from the intact tissue, that the rhythmic mechanism did not depend very strongly on internal calcium or on other parameters that may vary depending on the conditions of the isolation method. An alternative explanation demonstrated later in this article, is that the sinus node pacemaker mechanism is designed to be particularly resistant to conductance changes.

DiFrancesco, Ferroni, Mazzanti and Tromba (1986) succeeded in avoiding cell rounding and obtained what they called 'spider' cells, with several protuberances. Finally, Denyer and Brown (1987) succeeded in isolating single SA node cells which retain their natural thin spindle shape (Masson-Pévet et al, 1979). A recent series of papers (Denyer and Brown, 1990a, b, c) has used this method to describe both the basic electrophysiological properties and some of the mechanisms underlying rhythmic activity, including an attempt to assess the relative quantitative contributions of different inward currents to the generation of the pacemaker depolarization.

Nevertheless, a severe problem remained; this is that the patch clamping of such small cells inevitably leads to rapid run-down of conductance mechanisms. The calcium channels progressively disappear and the hyperpolarizing-activated current i_f shifts to progressively more negative activation ranges. This is the explanation for the wide variation in the activation range measured for this current (see, for example, DiFrancesco and Noble, 1989, Figure 2, in which the i_f activation curve varies over a range of about 35 mV in the position of the voltage for half activation). This is a particularly severe problem when it comes to assessing the contribution of i_f to the pacemaker depolarization since this depends entirely on the position of the activation range. Even a shift of 10 mV in this range can lead to a change in contribution

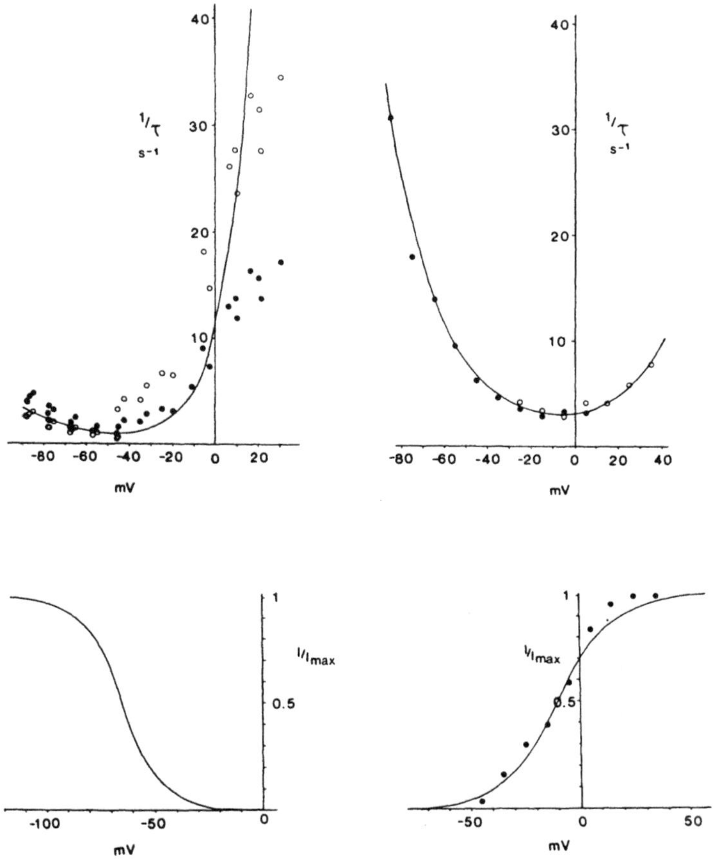

Figure 1. Kinetics of activation of i_K and i_f used in single sinus cell model. Left: the top diagram shows the reciprocal time constants of current change for i_f. The continuous line is computed from the functions used in the model. The bottom diagram shows the function used for the activation curve. Right: similar results for i_K (from Nobel, DiFrancesco & Denyer, 1989).

from this current mechanism of more than 100%.

Very recently, a method has been developed that permits patch electrodes to be used for clamping the whole cell without breaking the patch seal. Introduced by Horn and Marty (1988), it uses nystatin in the patch electrode to generate sufficient nystatin-induced conductance in the patch membrane to access the cell interior electrically without allowing the internal cell contents to be

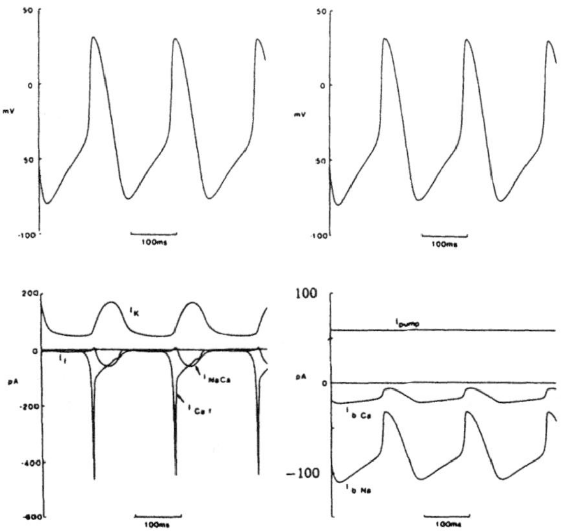

Figure 2. Computed pacemaker activity and ionic current using the single SA node cell model. Top: time course of computed voltage changes. Bottom left: time course of gated currents. Bottom right: time course of computed changes in background and pump currents (from Noble, DiFrancesco & Denyer, 1989).

dialyzed by the patch electrode.

This is the method that has been used to obtain the experimental data on which the latest computer model of single sinus node cells has been constructed (Noble, Denyer, Brown and DiFrancesco, 1992).

There are now, therefore, several versions of a mathematical model of the SA node using the approach first adopted by DiFrencesco and Noble (1985), including the original multicellular version (Noble and Noble 1984), its extension to a single cell by Noble, DiFrancesco and Denyer (1989) and the subsequent refinements described by Noble, DiFrancesco and Denyer (1991) and by Noble, Denyer, Brown and DiFrancesco (1992). There is also a very early model developed by Yanagihara, Noma and Irisawa (1980) which was the first SA node model, but this does not include concentration changes. We have extensively studied the properties of these models as a prelude to using them in very large network models and some important general conclusions have emerged.

Relative Roles of i_f and Inward Background Currents

This question is both important and still not fully resolved. It is also the question that has strongly motivated further refinements of the SA node cell models. The problem can best be presented by referring to Figure 1 which shows some of the experimental data on which the Noble-DiFrancesco-Denyer model was based. The data is shown as symbols, while the mathematical functions used are represented by filled lines. The data on the delayed K current is clearly fitted very accurately indeed, whereas the data on the hyperpolarizing-activated current, i_f, is fitted poorly. This is in part because the activation of this current is not strictly exponential (DiFrancesco, 1984), but the more important difficulty arises from the fact that the position of the activation curve varies between experiments. As noted already above, this range can be up to 30 mV in patch clamped cells. This is partly what motivated the recent experiments using the nystatin patch method. In the Nobel-DiFrancesco-Denyer model we simply used the mean position of the activation curve.

Figure 2 shows the resulting computed pacemaker activity together with the variations in ionic currents underlying this activity. It can be seen that the activation of i_f is fairly small; its peak is only -25 pA compared to an estimated -100 pA for the background current. In this model therefore, $i_{b,Na}$ contributes 4 times as much depolarizing current as does i_f. This is reflected in the fact that, when i_f is reduced to zero in the model (Figure 3, top) the change in frequency is relatively small. This is also observed experimentally (Denyer and Brown, 1990b) when i_f is blocked with caesium ions.

It might be thought that this combination of experimental and computed results settles the issue. Figure 3 (middle and bottom) shows why this would be a premature conclusion. In this figure we have performed the converse "experiment", i.e. we have progressively blocked the background current, $i_{b,Na}$, instead of blocking i_f. Since this current carries 75% of the depolarizing current during the

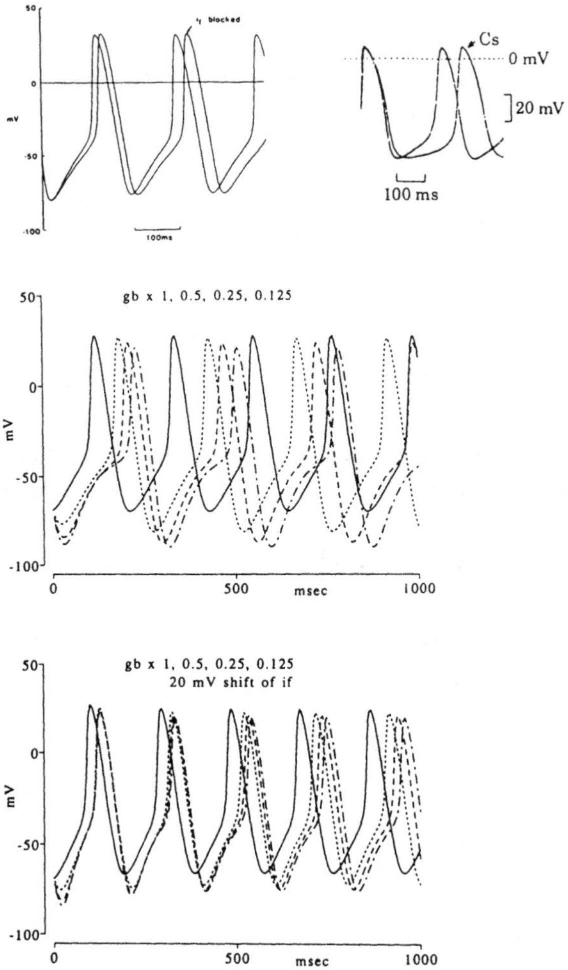

Figure 3. Top left: effect of block of i_f on computed pacemaker activity in the single SA node cell model. Top right: slowing of sinus node cell rhythm by block of i_f using 1 mM caesium ions. Middle: effect of reducing inward background current in SA node model to 0.5, 0.25, 0.125 of normal value. These changes induce moderate changes in frequency. Bottom: same computation repeated with i_f activation curve shifted 20 mV. The changes in frequency are now very small even when the background conductance change is very large.

computed pacemaker depolarization, it might be anticipated that the changes in frequency on changing its amplitude would be massive. In fact, however, the changes are not much larger than those obtained by blocking i_f. The top set of curves shows pacemaker activity computed from the model using background conductance values equal to 1, 0.5, 0.25 and 0.125 of the normal value. There is always an initial delay due to the immediate hyperpolarization that occurs. After this delay, the rhythm settles down to frequencies only moderately smaller than the control level. Even when the conductance is reduced massively to 0.125 of its normal value, the frequency change is no greater than that observed with complete block of i_f in many cells. Only by reducing $i_{b,Na}$ further does a dramatic change occur; pacemaker activity then stops.

The bottom of Figure 3 shows that even this assymetry between the two mechanisms cannot be taken as certain. Here we have repeated the calculations with the i_f activation curve shifted by 20 mV. This is within the range of experimental variation using the standard patch technique. It is clear that now the variation in frequency is extremely small and there was no level of conductance (even zero) which would stop the rhythm. Moreover, this is very similar to the results obtained using the newer model (Noble, Denyer, Brown and DiFrancesco, 1992) based on the nystatin patch method, where the frequency shows a strong resistance to massive changes in background conductance and does not stop when the conductance is reduced to zero.

These results are surprising to say the least. What they mean is, first, that either i_f or $i_{b,Na}$ are fully capable alone of providing all the depolarizing current during the pacemaker potential up to the point at which the calcium current is activated. Second, they show that regardless of the quantitative contribution, that i_f actually makes to the normal depolarization (and this might vary between different regions of the SA node and in different biochemical circumstances), its kinetic properties are finely tuned for it to act as a buffer against frequency changes induced by other changes in conductance. This means that the sinus node pacemaker is extremely robust and resistant to factors that might arrest it. The survival importance of such a 'fail-safe' system is obvious.

For the network modelling, we need to reconstruct regional variations of intrinsic cell properties (as seen in central,

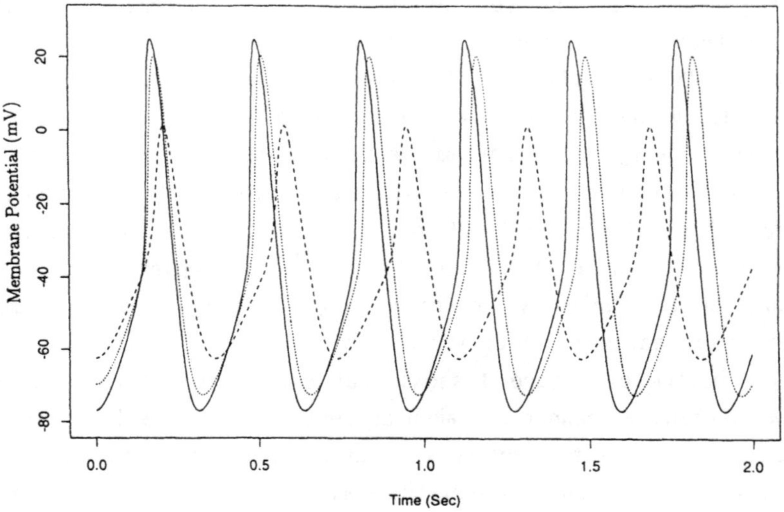

Figure 4. Computed variations of membrane potential with time for model peripheral (solid line), transitional (dotted line) and central (dashed line) cells (see Winslow et al, 1992 for details).

transitional and peripheral sinus node cells) such as frequency, amplitudes of action potentials and maximum diastolic potentials, upstroke velocity and sensitivity to external [K^+]. We have done this by adjusting the cell models to fit the data of Kodama and Boyett (1985). This information was obtained by segmenting a strip of the rabbit sinoatrial node running from its center to the periphery and then recording the electrophysiological characteristics in each isolated segment. We have found it possible by fine tuning of the individual channel conductances to reproduce Kodama and Boyett's data fairly well (Winslow et al, 1992). Figure 4 shows the computed records for cells representing the central, transitional and peripheral regions of the node. As one moves from the center towards the periphery, the maximum diastolic potential becomes more negative, the pacemaker depolarization becomes more rapid, the action potential amplitude increases and the upstroke velocity increases.

The first atrial model was developed from a sinus node model by Hilgemann and Noble (1987), and was used successfully to reconstruct

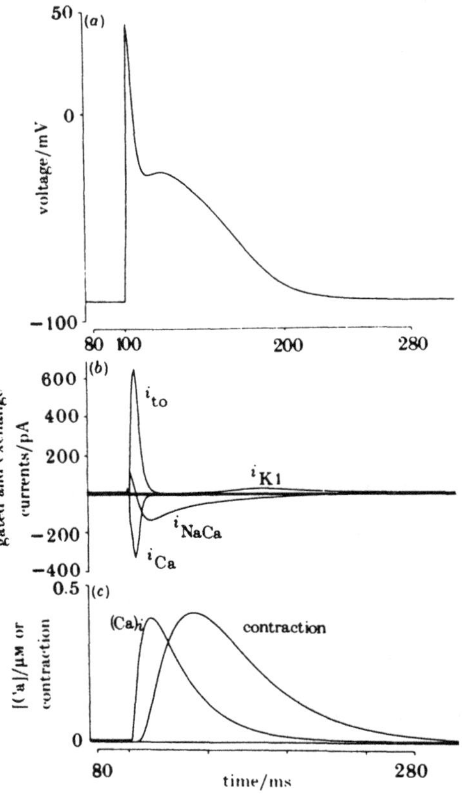

Figure 5. Atrial cell model: Top: action potential in response to an applied current stimulus. Middle: variations in the ionic currents k_{Ca} (calcium current), I_{NaCa} (sodium-calcium exchange current), i_{K1} (inward K rectifier current) and i_{to} (transient outward potassium current) (from Earm & Noble, 1990).

the movements of calcium during the heartbeat. Since one of the aims was to reconstruct fast extracellular calcium transients, the model was a multicellular one including an extracellular space. More recently, Earm and Noble (1990) have converted this into a single cell model using patch clamp data on isolated single rabbit atrial cells (Earm, Ho & So, 1990, 1991). Figure 5 shows the single cell model. Unlike sinus cells, there is no pacemaker depolarization.

The model requires a stimulus to respond with an action potential.

Network Modelling

Although there are still some details to be developed, the models of single cells are sufficiently accurate to enable us to have some confidence in the idea that they could now form the basis of the ultimate goal of trying to put the heart back together again. We have therefore incorporated the models of the single rabbit sinus node and atrial cells into large scale network models. These models are represented as N x N meshes with neighbouring cells electrically coupled by resistors representing gap junctions.

In the parallel computer computations, networks with as many as 1024 x 1024 cells are partitioned into sub-grids, mapping the set of differential equations defining each sub-grid onto different processors of a 32,768 processor Connection Machine CM-2 massively parallel processor (Thinking Machines Corporation), and integrating the equations of each sub-grid concurrently.

The first calculations were designed to determine the degree of cell-to-cell connection required to enable the sinus node to act as a synchronous unit and to send a coordinated signal to the atrium. This question was last tackled by one of us in 1982 (Noble, 1982) when a calculation based on using the multicellular voltage clamp data suggested that sinus node cells with different intrinsic properties could be synchronized with only a very few nexus channel connections between them. This calculation suffered however, from the assumption already referred to above, i.e. that the great majority of the cells in the multicellular preparations were behaving normally. This is unlikely, and if only a small fraction of the cells in the multicellular preparations were acting normally then the calculated individual cell conductances would have been divided into fewer units. Since we now know the single cell conductances very accurately, the question can now be answered definitively.

To achieve this, we have arranged 128 x 128 sinus cells into a network on the Connection Machine and initially, we have assumed random distribution of intrinsic properties. The nexus conductance between cells was then progressively increased from 2 channels to

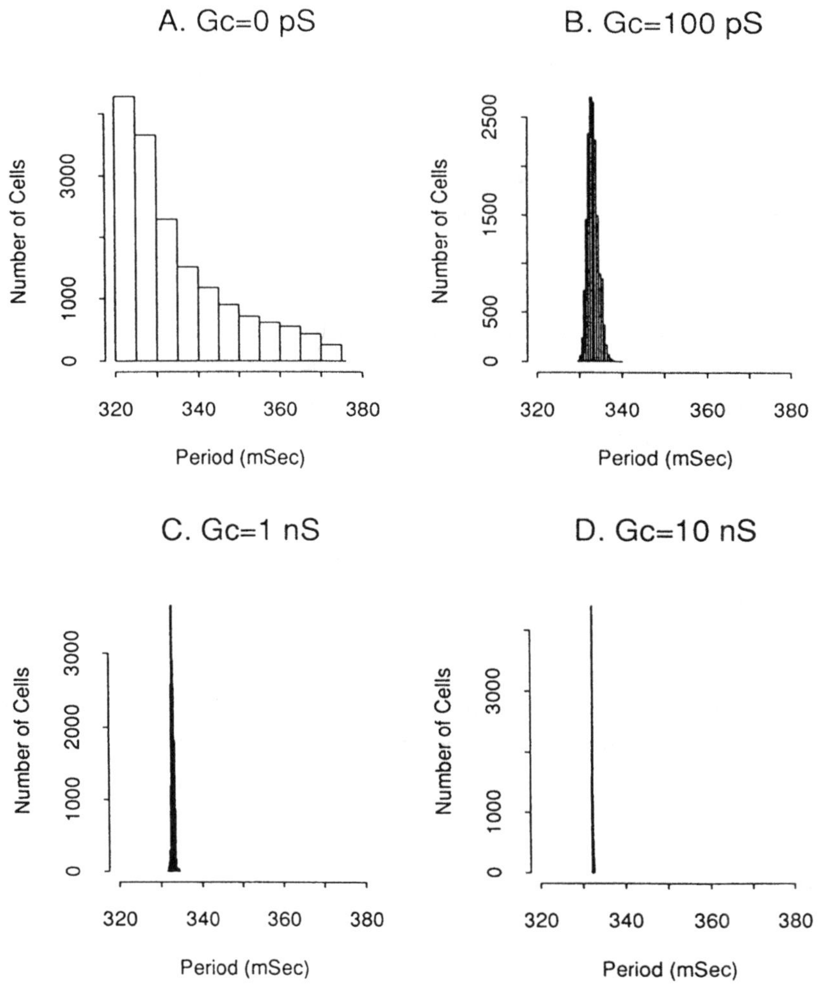

Figure 6. Oscillation period density functions obtained for sinus node cell network with random distribution of cell types. The plots show number of cells per period bin (ordinate) versus oscillation period (abscissa, msec). The connection conductances used in each calculation are A: 0 pS, B: 100 pS, C: 1000 pS, D: 10000 pS (from Winslow et al, 1992).

2000 channels (of 50 pS each) between cell neighbours. As shown in Figure 6, the results show that as few as two nexus channels connecting neighbouring cells can achieve considerable frequency entrainment, while 20 channels achieve almost complete synchrony. The original conclusion of Noble's (1982) calculation is still therefore valid. These figures correspond to a very low density of gap junctions, which would occupy a very small fraction indeed of the cell membrane surface. This result is consistent with the experimental observations (Masson-Pévet et al, 1979), showing that in the central region of the sinus node nexus regions are indeed very sparse and occupy less than 0.2% of the cell surface.

The second question we have tackled with the network modelling is how the wave of excitation propagates when the intrinsic single cell properties (fitted to the Kodama and Boyett data) are distributed between the node center and periphery using a Gaussian function. With the same magnitude of cell-to-cell coupling as in the random network, the results show that an excitatory wave then starts in the peripheral regions of the node and propagates towards the center, i.e. in the opposite direction to that in the normal heart. This result is very encouraging since this is exactly what occurs experimentally when the rabbit sinus node is separated from the atrium (Kirchhoff et al, 1987; Kirchhoff, 1989). Moreover, when the simulated sinus node is surrounded by an atrial network, the site of origin of the impulse shifts towards the center of the node (Winslow et al, 1992). The network modelling is therefore already achieving significant results in line with experimental evidence. Our intention in the future is to extend the modelling to arrhythmic mechanisms so that we can study the way in which cellular and multicellular mechanisms of arrhythmia interact.

Acknowledgments

Supported by the Army High performance Computing Research Center, The University of Minnesota, Minneapolis, MN, Thinking Machines Corporation, Cambridge, MA, the BHF, MRC and Wellcome Trust.

References

1. Bleeker WM, MacKaay AJC, Masson-Pévet M, Bouman LN, Becker AE. Functional and morphological organization of the rabbit sinus node. Circulation Research 1980;46:11-22.

2. Denyer JC, Brown HF. A method for isolating rabbit sinoatrial node cells which maintains their natural shape. Japanese Journal of Physiology 1987;37:963-965.
3. Denyer JC, Brown HF. Rabbit sinoatrial node cells: isolation and electrophysiological properties. Journal of Physiology 1990a;428:405-424.
4. Denyer JC, Brown HF. Pacemaking in rabbit isolated sino-atrial node cells during Cs^+ block of the hyperpolarizing-activated current i_f. Journal of Physiology 1990b;429:401-409.
5. Denyer JC, Brown HF. Calcium 'window' current in rabbit isolated sino-atrial node cells. Journal of Physiology 1990c;429:21P.
6. DiFrancesco D. Characterization of the pacemaker current kinectics in calf Purkinje fibres. Journal of Physiology 1984;348:341-367.
7. DiFrancesco D, Ferroni A, Mazzanti M, Tromba C. Properties of the hyperpolarizing-activated current (I_f) in cells isolated from the rabbit sino-atrial node. Journal of Physiology 1986;377:61-68.
8. DiFrancesco D, Noble D. Implications of the re-interpretation of i_{K2} for the modelling of the electrical activity of pacemaker tissues in the heart. In: Cardiac Rate and Rhythm (ed.) Bouman LN, Jongsma HJ. The Hague, Boston, London: Martinus Nijhoff 1982; pp. 93-128.
9. DiFrancesco D, Noble D. A model of cardiac electrical activity incorporating ionic pumps and concentration changes. Philosophical Transactions of the Royal Society 1985;B307:353-398.
10. DiFrancesco D, Noble D. The current i_f and its contrbution to cardiac pacemaking. In: Cellular and Neuronal Oscillators (ed.) JW Jacklet. New York: Dekker 1989;31-57.
11. Earm YE, Ho WK, So IS. Inward current generated by Na-Ca exchange during the action potential in single atrial cells of the rabbit. Proceedings of the Royal Society 1990;B240:61-81.
12. Earm YE, Ho WK, So IS. Effects of Ca^{2+}-buffer concentration and stimulus interval on the voltage dependence and timecourse

of calcium-release dependent inward current in rabbit atrial myocytes Proceedings of the Royal Society 1991; in press.

13. Earm YE, Noble D. A model of the single atrial cell: relation between calcium current and calcium release. Proceedings of the Royal Society 1990;240:83-96.
14. Hilgemann DW, Noble D. Excitation-contraction coupling and extracellular calcium transients in rabbit atrium: Reconstruction of basic cellular mechanisms. Proceedings of the Royal Society 1987;B230:163-205.
15. Horn R, Marty A. Muscarinic activation of ionic currents measured by a new whole-cell recording method. Journal of General Physiology 1988;92:145-159.
16. Irisawa H, Brown HF, Giles WR. Cardiac pacemaking in the sinoatrial node. Physiological Reviews 1992; in press.
17. Janse MJ, Wit AL. Electrophysiological mechanisms of ventricular arrhythmias resulting from myocardial ischemia and infarction. Physiological Reviews 1989;69:1049-1169.
18. Kirchhoff CJHJ. The sinus node and atrial fibrillation. Ph D Thesis, University of Maastricht 1989.
19. Kirchhoff CJHJ, Bonke FIM, Allessie MA, Lammers WJEP. The influence of the atrial myocardium on impulse formation in the rabbit sinus node. Pflügers Arch 1987;410:198-203.
20. Kodama I, Boyett MR. Regional differences in the electrical acitvity of the rabbit sinus node. Pflügers Arch 1985:404:214-226.
21. Masson-Pévet M, Bleeker WK, Mackaay AJC, Bouman LN. Sinus node and atrium cells from the rabgit heart: a quantitative electron microscopic description after electrophysiological localization. J Molec Cell Cardiol 1979;11:555-568.
22. Nichols CG, Lederer WJ. The regulation of ATP-sensitive K channel activity in intact and permeabilized rat ventricular myocytes. Journal of Physiology 1990;423:91-110.
23. Noble D. 1982; In discussion following De Haan RL. 1972; In: Cardiac Rate and Rhythm (ed.) Bouman LN, Jongsma HJ. Martinus Nijhoff: The Hague pp. 359-361.

24. Noble D. The Surprising Heart: A review of recent progress in cardiac electrophysiology. Journal of Physiology 1984;353:1-50.
25. Noble D. Ionic mechanisms determining the timing of ventricular repolarization: significance for cardiac arrhythmias. Annals of the new York Academy of Sciences 1991; in press.
26. Noble D, Denyer JC, Brown HF, DiFrancesco D. Modulation of pacemaker rhythm by conductance changes. 1992; in preparation.
27. Noble D, DiFrancesco D, Denyer JC. Ionic mechanisms in normal and abnormal cardiac pacemaker activity. Neuronal and Cellular Oscillators. J W Jacklet, Dekker, New York, 1989;59-85.
28. Noble D, DiFrancesco D, Denyer JC. Numerical reconstruction of the activation hypothesis. Journal of Physiology 1991; in press.
29. Noble D, Noble SJ. A model of SA node electrical activity using a modification of the DiFrancesco-Noble equations. Proceedings of the Royal Society 1984;B222:295-304.
30. Noble D, Noble SJ, Bett GCL, Earm YE, Ho WK, So IS. The role of sodium-calcium exchange during the cardiac action potential. Annals of the New York Academy of Sciences 1991; in press.
31. Noble D, Powell T. Electrophysiology of single cardiac cells. Academic Press 1987.
32. Noma A, Irisawa H. Membrane currents in rabbit sinoatrial node cells studied by the double microelectrode method. Pflügers Archiv 1976;364:45-52.
33. Tanaguchi J, Kokubun S, Noma A, Irisawa H. Spontaneously active cells isolated from the sino-atrial and atrio-ventricular nodes of the rabbit heart. Japanese Journal of Physiology 1981;31:547-558.
34. Winslow RL, Kimball A, Noble D, Denyer JC. Simulation of very large sinus node and atrial cell networks on the Connection Machine CM-2 massively parallel computer. Journal of Physiology 1991;438:180P.
35. Winslow RL. Modelling large SA node-atrial cell networks on a massively parallel computer. Journal of Physiology Proceedings of Cambridge meeting July 1991.

36. Winslow RL, Kimball AL, Varghese A, Noble D. Simulating cardiac sinus and atrial network dynamics on the connection machine. Physica D: Non-linear Phenomena, 1992; in press.
37. Yanagihara K, Noma A, Irisawa H. Reconstruction of sino-atrial node pacemaker potential based on the voltage clamp experiments. Japanese Journal of Physiology 1980;30:841-857.

Physiology of the Cardiac Electric Field

I. Ruttkay-Nedecky

Institute of Normal and Pathologic Physiology
Slovak Academy of Sciences, 81371 Bratislava, CZECHOSLOVAKIA

The Intracellular Electric Field

Production of electrical current and creation of electrical fields is implicit in all membrane functions that deal with electrolytes or transport coupled to electrolyte movement. Currents created by ion movements are not only a by-product of the transport process, but they are important biological events. There is indeed no reason to believe that potential gradients are confined exclusively to the outer membrane of the cell. More probably, they exist at each of the phase boundaries within the cell, at the nuclear membrane and at the surfaces of inclusion bodies for example.

It is also well known that enzymes are concentrated and operate preferentially at surfaces. Thus it was postulated by O.H. Schmitt that the chemical reaction mediated by an enzyme held at the surface occurs in the electrical gradient of that surface and in an orientation dictated by the geometrical alignments of the enzyme. In such a system, chemical energy released as a result of the enzyme mediated oxidation is not totally dissipated as heat, but becomes, to a large degree, available as electric energy at the interface. It is possible to imagine that such energy sources are mosaicked into biochemically differentiated areas, each specialized for the performance of an individual function (e.g. membrane channels, receptors).

The Transmembrane Action Potential

The living membrane structure depends for its integrity upon the continued existence of an electric gradient of a specific form in its vicinity. In turn, the generation of the gradient would certainly rely upon the integrity of the membrane structure. The underlying molecular processes that generate the cardiac electric field are connected with the activity of ionic channels and pumps at myocardial cell membranes.

Cardiac cells are able to change the permeability of their surface membrane to ions and thereby generate a stereotypical depolarization-repolarization sequence known as an action potential. At the membrane level, voltage differences generated by ion gradients influence ion transport by positive or negative feed-back. For example, sodium channels are gated to open or close by voltage. Both the sodium-potassium pumps and the sodium-calcium exchange system can generate a current large enough to affect in turn the action potential (1). The shape of the action potential is determined by the specific actions of voltage and time-dependent ionic permeability changes or conductances of the membrane.

The Extracellular Electric Field

An approaching excitation wave drives large currents out of the membrane by discharging the membrane capacitance and depolarizing the membrane to its threshold level. Past threshold, membrane has a large sodium conductance, so large inward currents flow through the membrane. This creates a source of current entering the extracellular space ahead of the advancing activation of the membrane, as well as a current sink just behind the excitation wave. The extracellular electric field reflects these currents and the positive and the negative components of the extracellular potential correspond roughly to the times of outward and inward net membrane current flow. Unipolar extracellular potentials recorded quite close to the cell membrane yield wave forms which resemble the second derivative of the intracellular action potential. Thus the second derivative approximation has been useful as a conceptual model for relating the upstroke of the membrane action potential and the potential curves recorded nearby.

The relationship between the intracellular action potentials and the extracellular wave forms are dependent upon the degree to which there is free movement of current in extracellular space. The duration of the biphasic extracellular curve approximates the duration of the upstroke of the action potential. The maximum of the extracellular curve occurs near, but preceding the apex of the upstroke of the intracellular curve. Where muscle fibers are packed together, the depolarization of the muscle produces a synchronous wave of extracellular activity which both precedes and trails the cells undergoing depolarization.

Extracellular waves were studied in vivo on the surface of a thin sheet of cardiac muscle on the dog vena cava superior. The distance between the maximum and the minimum of the wave varied between 0.6 and 1 mm (2). It has been pointed out (3) that results obtained by intramural recording of electric manifestation of the propagating activation front may be more conveniently explained by the assumption of the "oblique dipole layer" model, which represents the wave front as the sum of a normal dipole layer and an axial dipole layer and so takes into account the effect of myocardial anisotropy (4).

Propagation of Activation Wave Fronts

Large intracellular potential differences exist along the fiber between the partially depolarized region where the current is entering, and the fully polarized region ahead of the advancing excitation wave. These potential differences cause to flow intracellular axial currents, which in turn begin the process of capacitative discharge at the next portion of the membrane.

The transmission of the action potential from cell to cell usually occurs at the intercalated disk, which is a specialized region and contains large channels, called gap junctions, between cells. The gap junctions electrically couple cells by selectively allowing low-resistance passage of ions from cell to cell. Other electrical properties of these cells, such as membrane capacitance, myoplasmic resistance, and the interstitial resistance, affect also the propagation and generation of the action potential.

Since the myocardium is a functional syncythium, the propagation of excitation wave fronts occurs in all directions, but with a roughly three times higher velocity along the fiber axis than in a cross-fiber direction. As a consequence, the activation pattern arising from an ectopic stimulus is seen to be ellipsoidal.

While the direction of the plupart of cardiac bundle fibers is essentially tangential to endocardial and epicardial surfaces, isochrones recorded by plunge electrodes in the depth of ventricular walls show a clear initially radial propagation of activation wave fronts from the endocardium to the epicardium (5). Tangential propagation comes into play after breakthrough of the excitation wave on the epicardial surface of ventricles. This seemingly paradoxical pattern of ventricular activation is due to the indication of ventricular excitation along broad endocardial surfaces, from the network of subendocardial Purkynje fibers.

The junctional process between Purkynje and ventricular muscle cells is mediated by specialized transitional cells. There is a junctional delay of impulse propagation on the order of 3-5 ms (6,7). In addition to this temporal discontinuity, the junctional process is also spatially discontinuous, in that it occurs at some discrete sites of the endocardial wall.

There is apparently only a relatively small population of transitional cells which by their anatomical arrangement appear to be able to receive sufficient current for activation through high resistance coupling from the Purkynje cells. The action potential in the transitional cells then must supply the current for ventricular muscle cell activation across the transitional cell - ventricular cell junction of apparently low resistance. Thus the activation from the Purkynje fibre excites the transitional cells and their action potential becomes an active current generator for initiation of activation propagation in the working myocardial cells.

Critical analysis of practically all available cellular models of cardiac activation has shown that most likely activation of the myocardium takes a direction transverse to the fiber axis not only in the tissue but also at the cellular level (8).

Computer Simulation of Activation Propagation

The physiologically meaningful interpretation of body surface distribution depends on prior knowledge of myocardial activation propagation obtained laboriously by direct electrophysiological measurements (9). Recent computer simulation techniques allow to use effectively mathematical modeling for posing and answering questions which cannot be easily solved in a laboratory or in vivo situation. A recent critical review of mathematical models of cardiac activation is given in (10).

A Probabilistic Approach to Modelling

The concept of the cardiac electric field generator as a realistically localized multiple dipole, based on data from direct electrophysiological mapping of cardiac activation sequence, made it possible to give a forward start to the restricted inverse problem. The basic difficulty is how to introduce biologically realistic variability by means of inverse procedures into the inherently rigid parameters of the cardiac generator model, built up by means of forward procedures.

The concept of a sharply defined front of myocardial activation wave, moving at a uniform velocity is instrumental in clinical electrocardiology, as well as for description of its propagation. However, the velocity of excitation propagation depends on a variety of factors such as the curvature of the wave front, its size, the proximity of boundaries, the amount of connective tissue, the heart rate, etc. Also, the amount of the extracellular current generated by a unit wave front area is not uniform in different portions of a wave front, and even the direction of the current depends on the size of the wave front, local fiber direction, and the properties of the conducting medium (3).

Detailed studies have made clear that the activation front is irregularly shaped and that activation is not quite continuous (11). This may be caused by the irregular distribution of interconnections between the irregularly shaped myocardial cells.

In models of the cardiac electric field, the source is often conceived as an anatomically bound set of vectors, where the fixed origin of each vector represents a small area of the myocardium, and its magnitude and orientation the temporo-spatial evolution of the excitation front in this area.

At any instant of time t, throughout the cardiac excitatory cycle, we may define two mutually exclusive subsets E and R of the set of myocardial elements H, representing the excited and resting areas of the myocardium, respectively. The membership functions U_E or U_R for the element H_j with respect to subsets E or R may attain in the simulation of a unique depolarization the binary values 0 or 1 only.

When repeating such simulation experiments, it will be found, that a particular element, investigated at comparable instants of the excitatory cycle $H_{jt} \in H$ will not have the same values of membership function $u_E(h_{jt})$ and $u_R(h_{jt})$, respectively, since the propagation of the excitation front depends upon local ionic and metabolic conditions and these may undergo changes, e.g. under the influence hypoxia, local autonomic transmitter concentrations, heart rate, etc.

In order to characterize such a dynamic system, during a certain period of time, the subsets $\underset{\sim}{E}$ and $\underset{\sim}{R}$ may be regarded as fuzzy subsets E and R. The membership function of h_{jt} may then attain any value in the interval < 0.1 > and may be derived from the relative frequency of its activation during repeat measurements (12).

If the element h_{jt} is represented by a unit vector $\bar{v}_{jt}$ with membership function $u\bar{v}_{jt}$ so that its magnitude will be multiplied by the value of the membership function of h_{jt}, we will obtain a fuzzy subset of vectors $\underset{\sim}{\bar{V}}$ representing the beat-to-beat or intraindividual variability for a given time span of the elements of E.

After vectorial summation of the elements of $\underset{\sim}{\bar{V}}$ we will obtain a fuzzy resulting vector representing the activation front at time t for a given period of time.

Model Studies of Cardiac Activation

Our three dimensional model of cardiac activation (13) belongs to the family of models based on Huygen's principle which have been shown to be instrumental in elucidating the relation between ECG wave forms and vectorcardiographic loops on one hand, and physiological or pathological conditions of the heart on the other hand. A few examples of results from our simulation experiments are discussed below.

The ventricular walls of our model heart consist of a regular matrix of 162114 elements. Their shape is approximated by segments of 4 coaxial ellipsoids. The dimensions were obtained after fixation in distension from autopsied hearts with no evidence of disease (14). The model was effectuated with the aid of a three-dimensional integer field. The simulation of the myocardial activation sequence is a modification of the Huygens geometrical method of wave front construction in a three-dimensional isotropic and homogeneous region.

Normal anatomical variants of left bundle geometry were utilized (15) for determination of the starting points of ventricular muscle activation. The areas of ventricular wall activation fronts, determined by counting the model elements on simulated excitation boundaries, were found to be in fair relation to time-dependent changes of group means of the spatial vector magnitude over time function, obtained with the aid of two different, physically corrected, lead systems (16,17) from two samples of normal population (13).

The simulation of the normal anatomical variability of the left bundle ramification, derived from data published in (15), has shown that it may account for about 20% of the total normal interindividual variability of the 10 ms instantaneous QRS vector ("septal vector") in the sagittal projection and 13% in the horizontal plane projection (18).

Simulation of activation starting from one of the normal anatomical variants of left bundle ramification, characterized by slim and short anterior subdivision has shown that this variant may result in missing Q waves in leads I,aVL,V_{5-6}. It means that this specific ECG pattern, believed to be symptomatic of septal fibrosis (19), may be the result of normal anatomic variability of the left bundle ramification (20).

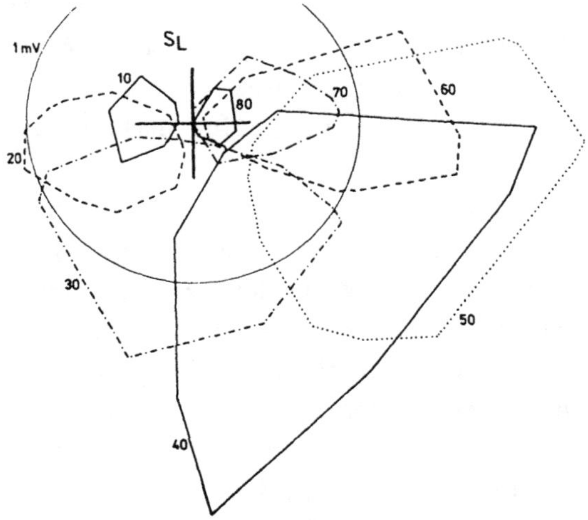

Fig. 1a). Left sagittal plane projection of vector termini at consecutive instants of QRS (10-80 ms). Spread of values after omitting 5% of marginal points is indicated by polygones. See text for further explanation.

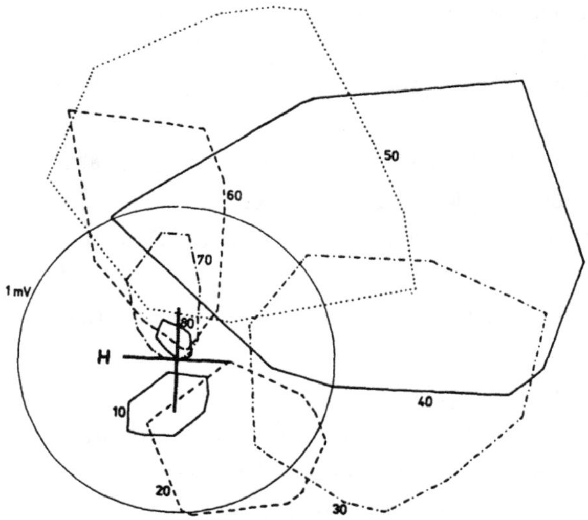

Fig. 1b). Horizontal plane projection.

Rotation of the model heart on its long axis so that the right ventricle was shifted anteriorly (clockwise rotation), resulted in a simulated vectorcardiographic pattern of right ventricular enlargement (21). Horizontal plane vectorcardiograms with signs of right ventricular hypertrophy were obtained by successively delaying right ventricular activation with respect to the onset of left ventricular activation (22).

Simulation of the effects of myocardial microinfarction have led to the conclusion that an electrically inactive area equivalent to about 9 mm^3 of myocardial tissue, might be detected as a transient decrease of the spatial magnitude over time function of instantaneous QRS vectors (20).

Physiological Evaluation of Surface Potentials

The objective of physiological evaluation of body surface manifestations of the cardiac electric field is to relate the potential distribution to electrical activity within the heart in a precise and predictable manner. While this end-point is still far away, important progress has been already achieved.

The sources of the cardiac electric field are localized in and around myocardial cell membranes. However, the potentials produced on the body surface during cardiac activation are those generated by atrial and ventricular wave fronts. These activation fronts are the effective generator of the extracardiac electric field due to atrial and ventricular depolarization. During repolarization, the current responsible for the T wave originates from the interaction of excitation sequence with the variation in action-potential duration in different parts (endocardium-epicardium, apex-base) of ventricular walls.

Vectorcardiography

Introduction of the physically corrected orthogonal lead systems (16,17,23,24) has provided the physiologist with a tool well suited for studying the intra-and interindividual variability of the cardiac electric field. With the availability of three simultaneously recorded orthogonal leads, it becomes possible to display the recording in several ways.

In Fig. 1a)b) are shown left sagittal and horizontal plane projections of instantaneous vector termini at 10 ms intervals after QRS onset obtained in a group of 145 healthy subjects aged 11-72 years (81 men, 64 women). The location of vector end-point subspaces is related at 10 ms after QRS onset to initial septal activation, at 20 ms to right and initial left ventricular wall depolarization at 30-50 ms to the balance of right and left free wall activation at 60-80 ms to terminal activation of the posterior basal ventricular wall.

Comparison of the angular and spatial velocities of QRS vectors allowed to identify the period of ventricular activation when angular velocity is at its minimum while spatial velocity is at its maximum. This may be a marker of the displacement of centroids of activation fronts during the depolarization of the apical region and the adjacent free ventricular wall (26). Study of respiratory variability of QRS vectors has disclosed non-uniform change of spatial magnitude in the course of ventricular depolarization (27).

Orthogonal, physically corrected leads are instrumental in comparing interspecies differences, as well as ontogenetic changes of the vectorcardiogram (28,29). Species-tailored, physically corrected lead systems were constructed for dogs (16), macaques and baboons (30).

Dipolar Electrocardiotopography

In dipolar electrocardiotopography (DECARTO), the orthogonal ECG signals are transformed to represent the equivalent generator of the cardiac electric field as a uniform double layer with time varying size and localization on a spherical surface enclosing the heart (31). Ventricular activation is then represented by time series of maps of "activated elements" on the discretized spherical surface. In Fig. 2, are shown maps of consecutive instants of ventricular depolarization obtained from the same group of subjects (25), as well as in another sample of healthy subjects using the Frank lead system. The spherical surface was transformed into a rectangle so that the left and the right borders of the rectangle coincide with the meridian of the sphere, viewing the right axillary region of the

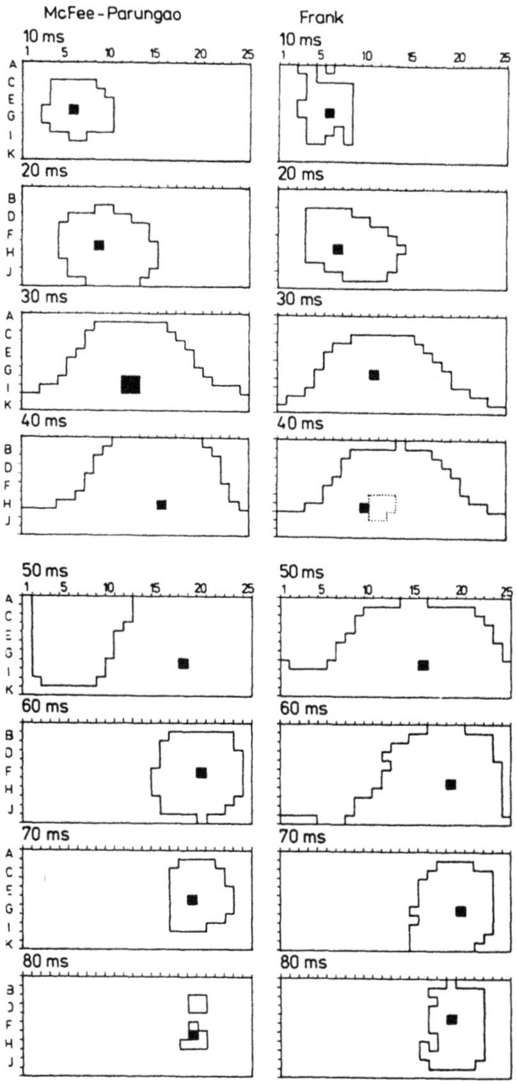

Fig. 2. DECARTO maps of ventricular activation at consecutive instants of QRS (10 ms intervals). The lateral borders of the map correspond to the right axillary midline. The left part of the map corresponds to anterior and the right part to posterior chest surfaces, respectively. Left panel: McFee Parungao lead system. Right panel: Frank lead system. Areas activated at 5% and higher probability levels are indicated by full lines. Black squares indicate areas of maximum probability.

subject. Limits of elements of the surface activated with 5% and higher probability are indicated by full lines. The localization of the maximum probability of activation for each distribution is marked by black squares. In this way of presentation, the areas of normal activation may be structuralized according to different levels of probability of activation of their elements.

Body Surface Potential Mapping

All the electrical information that can be elicited from the surface of the body is contained in maps of potential distribution obtained by a sufficient number of electrodes (up to 256). They provide more information on the electric activity of the heart than can be obtained from the standard 12 leads, or the orthogonal leads, because changes in the surface field may occur in areas which are not sampled by the aforementioned lead systems. On the other hand, the redundancy of surface potential mapping, as well as its sensitivity to extracardiac influences should also be taken into account.

Body surface potential maps have provided better insight into the sequence of cardiac events during excitation and recovery, particularly when two or more electrical processes simultaneously influenced the potential distribution. They have also provided new data on the duration of ventricular excitation and the overlapping of excitation and recovery potentials (33). They also allow to extract and to identify the "non-dipolar" information content of the cardiac electric field (34).

A most promising area of utilization of body surface potential mapping is the solution of the "inverse problem" of electrocardiology. With the knowledge of body surface and epicardial surface geometry, and by means of appropriate mathematical data treatment, it is possible to calculate potential distributions on the epicardium (35). The task is made easier if instead of epicardial surface, some geometric surface, enclosing the heart is utilized.

References

1. Fozzard HA, Friedlander JR. In: Comprehensive Electrocardiology I (eds.) PW. Macfarlane, TD Veitch Lawrie. Pergamon Press, New York 1989; pp. 79-100.
2. Spach MS, Barr RC. In: Advances in Electrocardiography (eds.) RC Schlant, J Willis Hurst. Grune and Stratton, New York 1972; pp. 19-36.
3. Taccardi BJ. Electrocardiology 1991;23:Suppl.150-154.
4. Colli-Franzone P, Guerri L, Viganotti C, Macchi E, Baruffi S, Spaggiari S, Tacaardi B. Circ Res 1982;51:330-346.
5. Durrer D, van Dam R, Freud GE, Janse MJ, Meijler FL, Arzbaecher RC. Circulation 1970;41:899-912.
6. Joyner RW, Picone J, Veenstra R, Rawling D. Circ Res 1983;53:526-534.
7. Joyner RW. In: Abstracts, 17th Internat. Congress on Electrocardiology, Florence, 1990; Sept. 26-29.
8. Plonsey R, van Oosterom A. J Electrocardiology 1991;24:99-112.
9. Van Dam RTh, Janse MJ. In: Comprehensive Electrocardiology I (eds.) PW Macfarlane, TD Veitch Lawrie. Pergamon Press, New York 1989; pp. 101-127.
10. Plonsey R, Barr RC. J Electrocardiology 1987;20:219-226.
11. Spach MS, Kootsey JM. Am J Physiol 1983;244:H3-22.
12. Ruttkay-Nedecky I. In: Le Coeur et l'Esprit (eds.) Reuse-Blom. Editions de l'Université de Bruxelles, Bruxelles, 1977; pp. 224-227.
13. Ruttkay-Nedecky I, Drkosová A, Szathmáry V. In: Advances in electrocardiology (ed.) P d'Alche. Centre de Publication de l'Université de Caen, Caen 1984; pp. 74-76.
14. Hutchins GM, Bulkley BH, Moore GW, Piasio MA, Lohr FT. Am J Cardiol 1978;41:646-654.
15. Demoulin JC, Kulbertus HC. In: Vectorcardiography 3 (eds.) I Hoffman, RI Hamby. North Holland Publ Co, Amsterdam 1976; pp. 123-127.
16. McFee R, Parungao A. Am Heart J 1961;62:93-100.
17. Rijlant P. J Physiologie 1960;52:267-322.

18. Ruttkay-Nedecky I. In: Progress in Electrocardiology. (ed.) PW Macfarlane. Pittman Medical, Tunbridge Wells 1979; pp. 115-117.
19. Burch GE, De Pasquale N. Am Heart J 1960;60:330-343.
20. Ruttkay-Nedecky I, Szathmáry V, Chlebus P, Ruttkay-Nedecká A. In: Models and Measurements of the Cardiac Electric Field (ed.) E Schubert. Plenum Press, New York 1982; pp. 35-41.
21. Szathmáry V, Ruttkay-Nedecky I, Eifrig Th, Sajban S. In: Electrocardiology '88 (ed.) H Abel. Elsevier, Amsterdam 1988; pp. 299-301.
22. Szathmáry V, Ruttkay-Nedecky I. In: Advances in Electrocardiology (eds.) Z. Antalóczy, I Préda, E Kékes. Elsevier, Amsterdam 1990; pp. 51-54.
23. Frank E. Circulation 1966;13:737-749.
24. Nelson CV, Gastonguay PR, Wilkinson AF, Voukydis PC. In: Vectorcardiography 2 (eds.) I Hoffman, RI Hamby, E Glassman. North Holland Publ Co, Amsterdam 1971; pp. 85-97.
25. Ruttkay-Nedecky I, Ruttkay-Nedecky P. Bratisl Lek Listy 1982; 77:403-412.
26. Ruttkay-Nedecky I, Rijlant P. In: The Electrical Field of the Heart (eds.) P Rijlant, I Ruttkay-Nedecky, E Schubert. Presses Académiques Européennes, Bruxelles 1972; pp. 468-471.
27. Ruttkay-Nedecky I. In: The theoretical bases of electrocardiology (eds.) CV Nelson, DB Geselowitz. Clarendon Press, Oxford 1976; pp. 120-134.
28. Ruttkay-Nedecky I, Cherkovich GM. The Orthogonal Electrocardiogram and Vectorcardiogram of Baboons and Macaques. Publ House of the Slovak Academy of Sciences, Bratislava 1977; pp. 61.
29. Diez U, Schwartze H. J Electrocardiology 1991;24:53-62.
30. Szathmáry V, Cherkovich GM, Ruttkay-Nedecky I. Physiologia Bohemoslovaca 1985;34:85-93.
31. Titomir LI, Ruttkay-Nedecky I. Int J Bio-Medical Computing 1987;20:275-282.
32. Ruttkay-Nedecky I, Titomir LI, Baum OV, Bacharova L. In: Electrocardiology '87 (ed.) E. Schubert. Akademie-Verlag, Berlin, 1988; pp. 145-148.

33. Taccardi B, De Ambroggi L, Viganotti C. In: The theoretical basis of electrocardiology (eds.) CV Nelson, DB Geselowitz. Clarendon Press, Oxford 1976; pp. 436-466.
34. Drska Z, Valová D, Málková A. In: Electrocardiology '87 (ed.) E Schubert. Akademie Verlag, Berlin 1987; pp. 165-167.
35. Horacek BM, de Boer RG, Leon JT, Montague TJ. In: Advances in body surface potential mapping (eds.) K Yamada, K Harumi, T Musha. The University of Nagoya Press, Nagoya 1983; pp. 47-54.

Modelling of the Cardiac Electrical Field

P. d'Alche

Laboratory of Animal Physiology and Bioinformatics,
University of Caen, 14000 Caen, FRANCE

WALLER, in 1889, was the first to record the electrical activity of the human heart with the capillary electrometer constructed by LIPPMANN. From 10 to 20 electrocardiograms recorded on a human chest, he established a crude body potential map. He observed that the potential distribution is analogous to that given by a single dipole with an orientation parallel to the base-apex cardiac axis. Thus the equivalent dipole generator was the first cardiac electrical field model. This concept forms the basis of classical electrocardiography and vectorcardiography. The single moving dipole representation and the multiple generator are refinements of the single fixed dipole. With these synthetic approaches, it is impossible to separate the activities of different parts of the heart. An analytical approach requires the use of multiple dipoles associated with numerous subdivisions of the heart. Such models describe the source of strength of cardiac cells. If the geometry, the heterogeneity and anisotropy of body tissues are taken into account, there are no fundamental limits to the complexity of this type of model.

The determination of an equivalent cardiac generator - such as a single fixed dipole, a single moving dipole or a multiple - from the surface potentials is known in electrocardiology as the "inverse problem". The calculation of surface potentials from a known distribution of electromotive forces is called the "forward problem". Multiple dipole and dipole layer models belong to the latter category. The development in the late 1950s of fast digital computers with large memories has made possible the numerical computing of solutions to the forward or inverse problems for realistic models.

This chapter is devoted to the description of the main cardiac source models, a discussion of the limits in modelling the cardiac electrical field and gives some examples of model utilizations.

Different Types of Generators Used in Modelling the Cardiac Electrical Field (Table 1).

Single Fixed Dipole (SFD).

In the heart as a whole, there are some millions of simultaneously active cells. The electrical activity of each cell may be depicted at each instant by a simple dipole. All these cellular dipoles can be combined into one single net resultant heart dipole. This approximation is still a common practice in electrocardiology. EINTHOVEN, in 1912, represented the human body as a plane, behaving like a homogeneous surface limited by a boundary defined by an equilateral triangle. In the centre of the triangle, he located the resultant heart dipole. Thus the origin of the resultant heart dipole is fixed but its direction, orientation and moment change at each instant of the cardiac systole.

A dipole is geometrically represented by a vector. Its direction and orientation are indicated by an arrow going from the negative to the positive charges of the dipole. The length of the arrow expresses the electrical moment of the dipole. Two non parallel electrocardiographic leads in the frontal plane were sufficient to define the frontal projection of the heart vector. Adding a third lead, outside the frontal plane, enables the spatial magnitude and orientation of the heart vector to be computed. LEWIS considered the electrocardiogram to represent a series of "heart vectors" and WILLIAMS in 1914 was the first to construct a vectorcardiogram which is a spatial curve connecting all the extremities of heart vectors throughout a cardiac cycle.

With multiple sources of current, the validity of the dipolar hypothesis depends on the distance from which the potential is

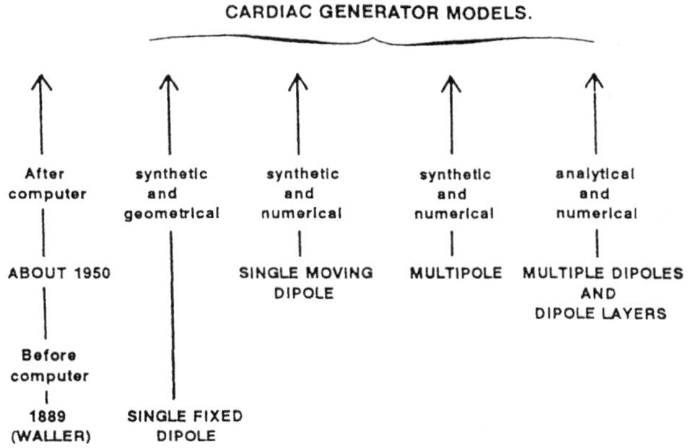

Table 1.

measured. The potential computed at any point outside the sphere including all the electrical charges is defined by an equation called multipole expansion characterized by different inverse powers of r (where r is the distance from origin to the observation point). Since bioelectric current sources have a zero algebraic sum, the zeroth multipole or monopole term does not occur and the first term including $1/r^2$ is the dipole term. The component including $1/r^3$ is the quadrupole term, $1/r^4$ the octupole term and so on. When the observation point is at a large distance compared with the overall source, then the higher terms are generally negligible.

As TACCARDI showed in 1958 (1) with an isolated turtle heart centered in a cylindrical conducting medium, the distribution of the isopotential lines becomes more regular as the distance from the cardiac surface increases. At a distance from the centre amounting to five times the radius of the heart, electrical effects are observed which are similar at any instant to those which would be generated by a single dipole.

In order to obtain an enlargement of conductive medium, RIJLANT suggested, in 1959 (2), the utilization of a resistive network. Using this network, he explored the entire surface of a human torso with 72 electrodes and in the same way the surface of an artificial body including an artificial generator made of six dipoles placed along three axes which are mutually perpendicular. The network is fed in spatial opposition by both sets of potentials measured in man and in artificial body so that they cancel out. With this physical simulation, RIJLANT could evaluate the quantitative characteristics of the cardiac generator, in particular the spatial orientation and magnitude of the heart vectors.

Single Moving Dipole (SMD).

In this model, the dipole location is allowed to move. The origin location of the dipole or "electrical center" of the heart is achieved by two methods. The first method uses the multipole expansion. The origin of the multipole expansion is shifted in order to minimize the coefficients of the quadrupolar term. A practical approach was given by GABOR and NELSON in their classic paper (3). In 1977, GESELOWITZ (4) proposed an alternative solution of the first method where the contribution of the quadrupole component to the surface potentials was smallest. The underlying principle of the second method is to find the particular dipole moment and location that best fit the experimental and estimated surface potentials in a least-squares error. One of the earliest approaches of this type was proposed by HELM and CHOU in 1971 (5).

Numerous authors have applied the SMD to the study of normal or pathological cardiac cycles in human or animal subjects. Thus in simulation and experimental studies, SAVARD et al, (6) have shown that the trajectory of the dipole during QRS can portray the passage of an ectopic beat across the heart.

Multipole

Closely related to the SMD generator is the multipole generator (YEH et al, in 1958) (7). As this has already been described (cf. single dipole generator), all the cardiac electrical charges may be represented by a multipole expansion. The coefficients for the dipole, quadrupole and eventually octupole terms are obtained by evaluating the potential distribution over the torso surface with appropriate geometrical weighting. From a practical standpoint the multipole representation allows a quantitative description of the nondipolar contribution to the surface potential field.

Multiple Dipoles and Dipole Layers.

In this analytical approach, the heart is divided into multiple polyhedral elements which may vary from fewer than 10 to several hundred thousand. Each element is assumed to have an electrical activity represented by a dipole. The analytical models may or may not include automatic propagation of activation from the conduction system. The moment of the dipoles may be defined at each instant from the assigned stylized action potentials and there exist four distinct states of electrical activity for each element: at rest, activated, in absolutely refractory period or in relatively refractory period. With such conditions, numerous models (8,9,10,11,..) have been proposed since the first models introduced by BARBER and FISCHMANN in 1961 (12), then SELVESTER et al in 1965 (13). Sometimes the activation isochrone fronts are represented by a uniform double layer of dipoles. The influence of body limits, of heterogeneity and anisotropy of different living tissues are or are not taken into account.

As an example, d'ALCHE et al, in 1974 (14), built the first model based on human depolarization data published by DURRER et al (15). The equations used were obtained from GREEN'S theorem with the assumption that at each point of an interface the current is normal to the surface. The potential Ø at point P is given by:

$$\Phi(P) = \frac{1}{4\pi\sigma}\int_v \frac{1}{R} dv + \frac{1}{4\pi\sigma}\sum_i \int_{S_i} (\sigma_i'' - \sigma_i')\,\Phi(r)\,\vec{\nabla}\left(\frac{1}{r}\right).d\vec{S_i}$$

Where Si is the surface separating two mediums with σ'_i et σ''_i conductivities. The epicardial and thoracic ECGs or potential maps obtained with the model are in good concordance with the actual records (16). The complexity of the field is easily explained if the model is used to separate electrical activity generated by different zones of the heart (Fig. 1). It is also possible to study the influence of the lungs on the distribution of thoracical potentials (17).

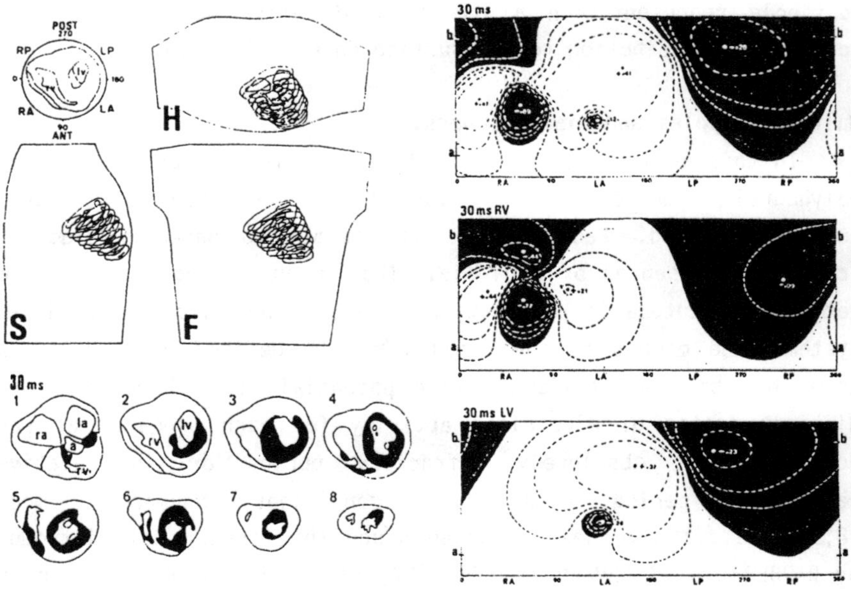

Fig. 1. Computer model of cardiac electrical activity built by d'ALCHE et al (14) based on human depolarization data published by DURRER et al (15). As an example, isopotential maps are computed on a cylindrical surface surrounding the heart from the activation surfaces, at 30 ms, located respectively in the two ventricles, in the right ventricle (RV) and the left ventricle (LV) together with the septum.

Discussion of Limits in Modelling the Cardiac Electrical Field.

In order to model the cardiac electrical field, many difficulties have to be overcome.

Difficulties in the Choice of Equations.

The elementary dipole strengths of depolarization wave fronts are not identical and the electrical conductivity of the living tissues is anisotropic. Therefore, the assumption of the uniform double layer which assumes that the dipole sources are of equal value and normal to the cell surface is not valid.

In the axial theory proposed by CORBIN and SCHER, in 1977 (18), all elementary dipoles point in the fiber direction. Their strength is proportional to the cosine of the angle between the direction of

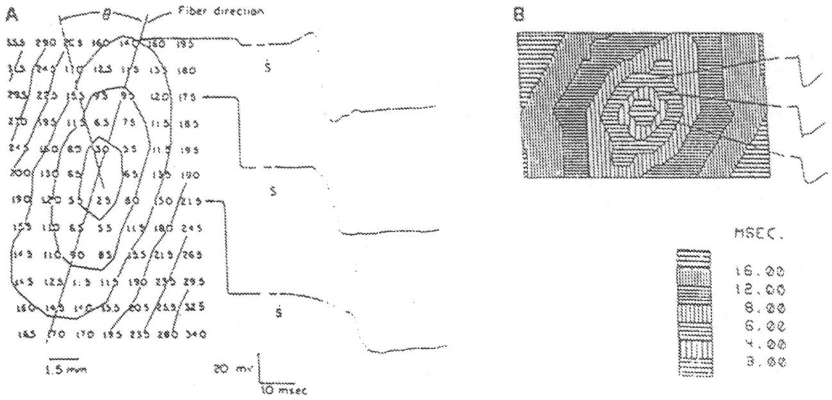

Fig. 2. In A, times of activity (msec) on the canine ventricular epicardium for the spread of the depolarization wave away from a central stimulation point. Three waveforms, on the right, were recorded by ROBERTS et al (21). In B, simulated activation and same waveforms as previously computed with the OUMELLAL and d'ALCHE model (20).

propagation of the wavefront and that of the fiber. Another formulation was given by COLLI-FRANZONE et al, in 1982 (19), called the "oblique dipole model": in addition to the axially oriented sources of the axial theory, dipole sources are postulated that are oriented in the transverse direction.

Using different velocity coefficients in relation to the direction of the cardiac fibers, OUMELLAL and d'ALCHE (20) computed the activation and potential maps at the surface of a simulated strand of myocardium. When the activation starts from a central element, the activation front is characterized by an ellipsoidal shape owing to faster propagation along the longitudinal cellular axis in the same manner as in the experiments performed by ROBERTS et al (21). Three waveforms in the same locations as those in ROBERT'S experiments were computed with the model. The actual and simulated waveforms are very similar (Fig. 2). Longitudinal waves show positive potentials before local depolarization and rapid intrinsic

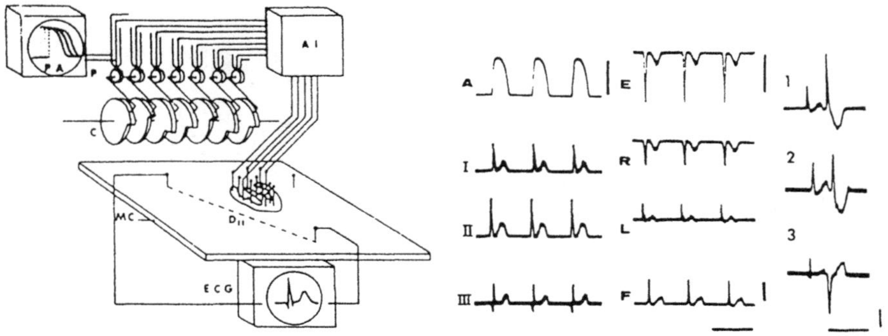

Fig. 3. Electro-mechanical model for reconstruction of normal or pathological ECGs from action potentials, d'ALCHE and CORABOEUF, 1967 (22).

deflections. Transverse waveforms show negative potentials before local depolarization and less rapid intrinsic deflections. The potential map shows positive potentials ahead of the wave along the fiber direction and negative potentials in the transverse direction, in accordance with actual results (20).

Difficulties in Integrating the Cellular Electrical Activities.

Any realistic model must integrate all the data playing a part in the genesis of the cardiac electrical field. For instance, the distribution of body surface cardiac potentials has to take into account cellular potentials recorded at the level sources. One of the first attempts in this direction was proposed by d'ALCHE and CORABOEUF, in 1967 (22) and 1971 (23), with an electro-mechanical model for the reconstruction of normal or pathological ECGs from action potentials (Fig. 3). For a more precise simulation, the kinetics of the membrane currents have to be included. A model of cardiac cellular activity was given by BEELER and REUTER in 1977 (24), as well as by DI FRANCESCO and NOBLE in 1985 (25). ROBERGE et al (26) used a network of interconnected BEELER-REUTER units to simulate a two dimensional sheet of cardiac tissue. However, the cellular electrophysiology of the heart is notoriously complicated.

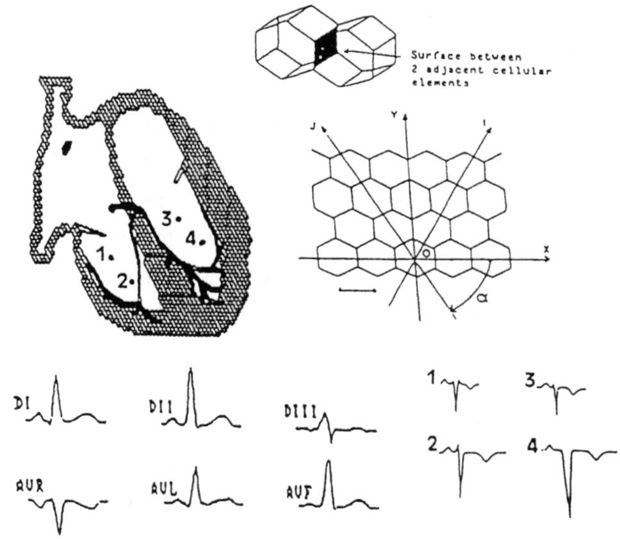

Fig. 4. Computer model built by OUMELLAL and d'ALCHE (20). Top right, cardiac diagram with the nodal and conductive tissues; left, geometry and network of cellular elements with their system of reference. Bottom, examples of simulated standard (DI, DII, DIII, aVR, aVL, aVF) and intracardiac electrocardiograms (1,2,3,4).

BARR and PLONSEY (27) estimated that to calculate one activation sequence from membrane currents in a two-dimensional model tesselated into a mesh of 10^6 nodes would take 3,000 years on an IBM 370/165 computer.

Imprecision and Variability of Anatomical and Physiological Data.

Many precise data must be included in a realistic model but the majority of these data concerning, for example, depolarization, refractory period duration, propagation velocity, fiber orientation, cardiac or thoracic anatomy, resistivity of different mediums and so on, are generally not known precisely. In spite of use of mapping systems such as the SATAPEC system built by d'ALCHE et al (28), allowing the simultaneous recording of up to 256 epicardiac or intraparietal electrograms, it is impossible at the present time to obtain very precise depolarization or repolarization maps for the whole heart. Furthermore, the anatomical and physiological data are not identical for each person or animal.

Difficulties in Assessing the Validity of a Cardiac Model.

A simplified model may provide a good enough simulation of electrocardiographic tracings or potential maps close to the actual ones. OUMELLAL and d'ALCHE (20) have demonstrated that it is relatively easy to simulate normal or pathological ECGs in spite of major simplifications (Fig. 4). In our model the heart is assumed to be placed in a homogenous and infinite medium. Only the part of the heart lying between two planes parallel to the frontal plane was considered in the computation. Such a model is, therefore, intermediate between 2D and 3D models. The auricles and ventricles have been cut into about 1000 polyhedral "cellular elements" of hexagonal section. Each "cellular element" is representative of a homogeneous cardiac area and is, at any given instant, in one of the following states: i) at rest, ii) activated, iii) in the refractory period. In the next instant, the activated "cellular elements" activate the surrounding elements at rest. The propagation of activation through the conductive tissue is in conformity with the classical data. The repolarization process depends on the duration of the activation of each "cellular element". The potential at one point is computed from each separate surface between activated and resting cellular elements by the solid angle formula, modified in order to take into account the velocity coefficients in the myocardium and the extracellular potential drop in relation to the direction of the activation and the supposed longitudinal axis of each "cellular element". Fig. 4 shows some simulated standard (DI, DII, DIII, aVR, aVL, aVF) and intracardiac electrocardiograms (1,2,3,4) obtained with the model. The amplitude and duration of each tracing are in the same arbitrary units. In most cases, the simulated tracings are very similar to the actual ones (the similarity has been evaluated by a correlation coefficient).

Examples of Applications of Modelling the Cardiac Electrical Field.

The more valuable applications of cardiac models concern the

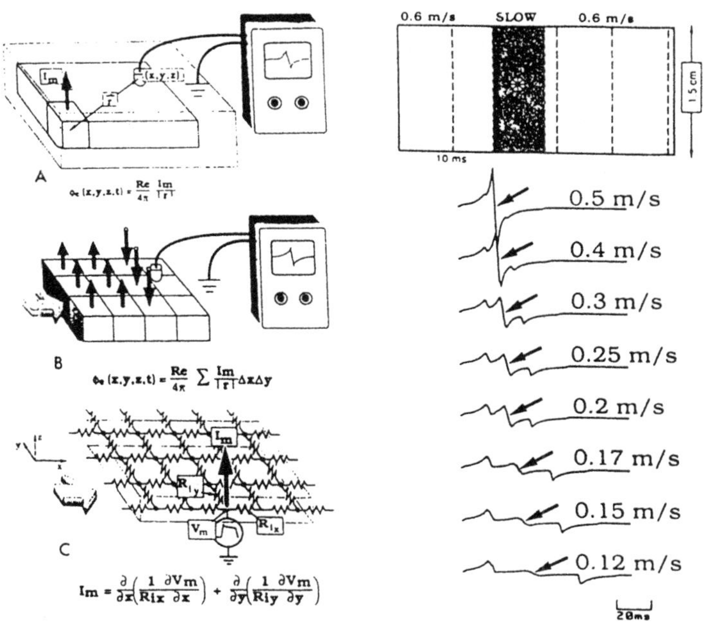

Fig. 5. Computer model of electrograms built by LESH et al (29) with tracings, recorded from central electrode, showing varying degrees of slow conduction.

interpretation of electrocardiographic tracings or their use as a research tool.

Interpretation of Electrocardiography Tracings.

Fractionated electrograms are frequently recorded by endocardial catheter mapping in patients with coronary artery disease and ventricular tachycardia. What causes fractionation? In 1988, LESH et al (29) developed a computer model of electrogram generation which is a mathematical transposition on a cardiac strip of the electrical model for the whole heart built by d'ALCHE and CORABOEUF in 1967 (22). The model consists of 5,000 elements in a two-dimensional grid simulating a 30 x 15 x 0, 1 mm area of ventricular myocardium. Each element is coupled to its neighbors through a resistance (Fig. 5). Activation time of each element depends on its position in the grid and the conduction velocity. Each element utilizes the same action potential once activated. The action potential voltage Vm is related

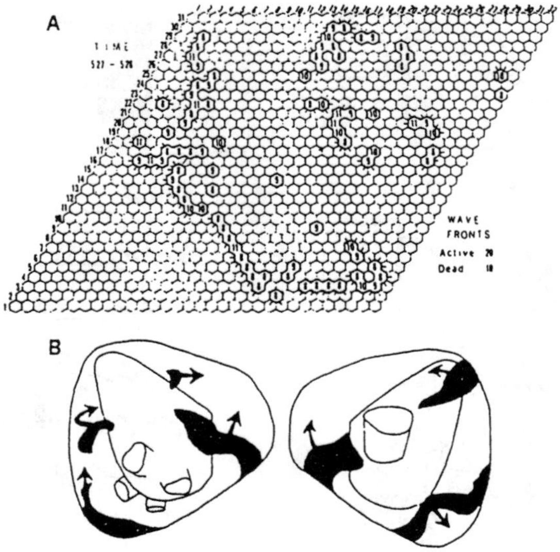

Fig. 6. In A, wave fronts during atrial fibrillation computed with the model developed by MOE et al (30). In B, multiple propagating wavelets in canine atria encountered during atrial fibrillation as recorded by ALLESSIE et al (31).

to the transmembrane current Im by:

$$Im = \delta/\delta x\ (1/R_{ix}.\delta Vm/\delta x) + \delta/\delta y\ (1/R_{iy}.\delta Vm/\delta y)$$

x, y, z give the electrode location in space, R_{ix} and R_{iy} are the interelement resistivity along the x and y axes.

The simulated results are validated by checking them against electrograms obtained from epicardial strips of tissue excised from mongrel dogs and superfused with oxygenated Tyrode's solution. After recording unipolar electrograms in the control state, a zone of slow conduction is produced by painting the central portion of the strip of tissue with heptanol.

LESH et al prove that intracellular resistivity changes were responsible for the polyphasic nature of the electrogram while conduction velocity changes were responsible for increased electrogram width. In tissue from which fractionated electrograms are recorded, local activation may not correspond to the largest or most rapid deflection in the local electrogram.

Cardiac Models as a Research Tool.

MOE et al, in 1964 (30), used a model for understanding the genesis of atrial fibrillation. At that time, the electronic equipment required for detailed analysis of the complex excitation pattern of the atria during fibrillation in an intact heart was not available. For this reason, MOE et al developed a mathematical computer model representing a piece of cardiac tissue in which atrial fibrillation could be simulated. They introduced a random spatial distribution of refractory period of the simulated cardiac cells. The application of a short burst of external stimuli to a small cluster of cells exhibiting short refractory periods resulted in a self-sustained arrhythmia resembling fibrillation. The spread of excitation in this model showed progressive disorganization with no fixed reentrant pathways; instead the circuits constantly died and were replaced by others resulting in a continuous shift of position, frequency and direction of circuits (Fig. 6A). These model studies have strengthened the probability of the multiple wavelet hypothesis as an explanation for the mechanisms underlying fibrillation of the heart.

The experimental proof of such a hypothesis was given by ALLESSIE et al (31) in their studies of atrial fibrillation in isolated Langendorff-perfused canine heart. From 192 unipolar electrograms recorded simultaneously by means of a mapping system, they drew by hand the isochrone lines of the atrial activation during the fibrillation. They observed that multiple wandering wavelets of different sizes and travelling in various directions were present during fibrillation. Their studies showed (Fig. 6B) that there exists a critical number of wandering wavelets for the perpetuation of fibrillation (between three and six).

In conclusion, at the present time, it is impossible to obtain all the precise data necessary for building a realistic model for complex applications such as aid to diagnosis or the computation of electrograms from thoracical potentials. The main difficulty is due to the fact that the anatomical and physiological data are not identical for each person or animal. However, a model of cardiac electrical activity may be considered as a research tool, the results of which have to be confirmed by experiments. A model may help to solve many simple problems such as the interpretation of electrocardiographic tracings or the genesis of arrhythmias. With technical progress, a new generation of models will probably appear in the future.

References

1. Taccardi B. Acta Cardiologica 1958;13:173.
2. Rijlant P. Bull Acad Roy Med Bel 1959;24:718-745.
3. Gabor D, Nelson CV. J Appl Physics 1954;25:413-416.
4. Geselowitz DB. IEEE Trans Biomed Eng 1977;24:242-252.
5. Helm RA, Chou TC. In: Vectorcardiography (eds.) I Hoffman, RI Hamby, E Glassman. North-Holland, Amsterdam 1971; pp. 98-106.
6. Savard P, Roberge FA, Perry JB, Nadeau RA. Circ Res 1980;46:415-425.
7. Yeh GCK, Martinek J, de Beaumont M. Bull Math Biophys 1958;20:203.
8. Ruttkay-Nedecky I, Drkosova A, Szathmary V. In: Cardiology (ed.) P. d'Alche. Univ Caen 1984; pp. 74-76.
9. Aoki M, Okamoto Y, Musha T, Harumi K. IEEE Trans Biom Eng 1987;34:454-462.
10. Gulrajani RM. Crit Rev Biomed Eng 1988;16:1.
11. Leon LJ, Horacek BM. J Electrocardiol 1991;24:1-43.
12. Barber MR, Fischmann EJ. Nature Lond 1961;192:141-142.
13. Selvester R, Collier CC, Pearson RB. Circulation 1965;31:45-53.
14. d'Alchá P, Ducimetiere P, Lacombe J. Circ Res 1974;34:719-729.
15. Durrer D, Van Dam RT, Freud GE, Janse MJ, Meijler FC, Arzbaecher RC. Circulation 1972;41:899-912.
16. d'Alché P. Adv in Cardiol 1976;16:47-51.
17. d'Alché P. In: Le coeur et l'esprit (ed.) S. Reuse-Blom. Univ Bruxelles, 1977; pp. 44-47.

18. Corbin LV, Scher AM. Circ Res 1977;41:58-67.
19. Colli-Franzone P, Guerri L, Viganotti C. Potential Circ Res 1982;51:330-46.
20. Oumellal H, d'Alché P. In: Electrocardiology 88 (ed.) H Abel. Excerpta Medica 1989; pp. 273-277.
21. Roberts DE, Hersh LT, Scher AM. Circ Res 1979;44:701-712.
22. d'Alché P, Coraboeuf E. CR Soc Biol 1967;161:30-34.
23. d'Alché P, Coraboeuf E. J Electrocardiol 1971;4:187-198.
24. Beeler GW, Reuter H. J Physiol (Lond) 1977;268:177-182.
25. DiFrancesco D, Noble D. Phil Trans Royal Soc B 1985;307:353-398.
26. Roberge FA, Vinet A, Victorri B. Circ Res 1986;58:461-672.
27. Barr RC, Plonsey R. Biophys J 1984;45:1191.
28. d'Alché P, Morel M, Gauthier V. Innov Techn Biol Med 1990;11:183-195.
29. Lesh MD, Spear JF, Simson MB. J Electrocardiol 1988;21:S69-S73.
30. Moe, GK, Rheinboldt WC, Abildskov JA. Am Heart J 1964;67:200-220.
31. Allessie M, Lammers W, Bonke F, Hollen J. In: Cardiac Electrophysiology and Arrhythmias (eds.) D Zipes, J Jalife. Grune et Stratton, 1984; pp. 265-275.

The Temporal Behaviour of the Cardiac Electric Field

E. Schubert, A. Patzak and W. Laube

Institute of Physiology, Humboldt-University, 0-1040 Berlin,
FEDERAL REPUBLIC OF GERMANY

Characterization of the Cardiac Electric Field.

The cardiac electric field represents a biological realization of a physically defined electrical field. Therefore its generation is performed by the electrical properties of a source which extends its efficiency over a conducting medium. The source is determined by its localization and strength. The conductor influences the resulting field by its conductivity and boundaries.

The role of the source in the case of the cardioelectric field is accomplished by the heart. Within its conducting syncytium of the myocardium, the heart possesses two parts of different electrical charge during the activation-recovery cycle. The still resting and the again fully recovered myocardial cells carry a positive charge on their outside. In contrast to that, the cells of the excited part are negatively charged on their surface. The charges of each of these regions can be concentrated into one geometrically well defined point. The confrontation of these two charged points forms the source of the field. The strength of the source is defined by the difference between the charges which normally amounts to about 100 mV. The geometric connecting line of the two charged points gives the expression of the localization in space.

The intellectual construction of the dipolar representation of the heart leads to the model of the source of the cardioelectric field named equivalent dipol. This dipol has frequently been utilized for considerations and constructions of problems of the field. In many cases it has proven as sufficient. For special questions mostly concerning precordial investigations and pathological cases of disturbed conduction pathways in the myocardium higher moments as quadrupols or octapols necessarily have to be stressed.

The conducting medium for the cardiac electric field is formed from the body tissues. As there exist large differences in the conductivity of the various organs surrounding the heart - muscle and blood are of high conductance, the lungs possess a low and complex conductance - the body has to be described as an inhomogeneous conductor. That influences the spread of the field in an irregular way. Therefore deformations in the field have to be expected. At the body surface a sudden change of conductivity of several orders of magnitude occurs between the tissues and the surrounding air. This practically produces a limitation of the field to the boundaries of the body surface and generates an imaging of the configuration of the field at the boundary. This can be recorded and gives the body surface field maps. Such a body surface map is the representation of the cardiac electric field of the electrical source situated in the heart. The conception of an electric field of the exciting heart has already been developed in the earliest days of electrocardiography by A.D. Waller in 1889 (1).

Examination of the Cardiac Electric Field.

The analysis of physical phenomena investigates usually two different qualities, the spatiality as the behaviour in space and the temporality as the reaction in time. Both approaches deliver various information concerning the structural and functional properties of the phenomenon under observation. These two entrances may lead to various methods of investigation.

The spatial configuration of the field characterizes its appearance on the body surface as well as in the tissues of the body. The description is given in the orthogonal Cartesian x y z - coordinates or in the polar coordinates of magnitude, azimuth and elevation. The real measurement is performed from the body surface and consists in the analysis of the distribution of the potentials on the surface producing the body surface map (BSM). The BSM reflects the configuration of the source regularly containing a region of maximum potential within a positive part of the field and a region of minimum potential within a negative part. The positivity projects the resting myocardium, the negativity the excited part of the heart (2). The correspondence between the excitational state of the heart and the configuration of the surface field is always visible, but the deforming influences of the inhomogeneous conducting tissues on the localization of the maximum and minimum and on the configuration of the field are obvious.

The comparison between the configuration of the field maps and the momentary state of the source allows to find spatial connections. This is done especially for the case of the localization of the axis of the heart and the spatial orientation of the R-vector. The first and easiest investigation of this topographic information out of the field was proposed by the procedure of the construction of the Einthoven's triangle (3). Starting from this construction topographic informations in different lead systems have been performed and utilized for the derivation of clinical diagnoses considering aberrations in localization or contribution of the source to the surface field. The most familiar example is the diagnostic of position anomalies of the heart or of partition variations of the source in the case of myocardial hypertrophies. As a result the spatial analysis of the cardiac electric field is directed towards the exploration of structural informations.

Introducing the course of time to the exploration of the field delivers aspects describing the function of the source. This way of investigation was introduced into physiology by C. Ludwig's inventions of the "Stromuhr" and the "Kymographion" in the middle of the last century.

The consideration of the excitatory cycle of the heart accomplished precise knowledge about the pathway of the activation front inside the myocardium that divides the resting from the activated part of the heart (4,5). Out of this the dynamics of the temporal changes in the cardiac electric field within one cycle can be derived. The information received in this way of investigation concern the time course of the excitatory process within the conducting pathway in the heart.

Also the mechanism of triggering the start of the cycle shows time related phenomena. The first observation results in the fact that the cardiac electric field in its complete time course of excitation is repeated nearly identically from cycle to cycle. This repetition has its causes in the rhythmical triggering of the development of the field i.e. the start of the excitatory cycle in the heart by the activity of the sinus node. Thus the regular series of heart beats is generated and its temporal behaviour becomes expressed in heart rate as the characteristic of heart rhythm.

Temporal structures within the rhythmical behaviour of heart rate may be revealed by more precise observations. Rough investigations discovered connections to the respiratory rhythm leading to the definition of respiratory arrhythmia. This is generated in the sinus node, therefore, the term sinus arrhythmia is also used. The application of modern methods of data processing has opened ways for an analysis of rhythmical substructures in the behaviour of heart rate. That helps in discriminating several components of the rhythm marked by different frequencies and participating in various intensities in the influencing the variability of heart rate. All these phenomena are transmitted by the control of the activity of the sinus node via the vegetative cardiac nerves. Therefore insights into the control of triggering and formation of the cardiac electric field are to be expected by such kinds of analyses of heart rate. Variability and its components constitute the rhythmicity and contain the total intensity of arrhythmia as one and the various components as another aspect of temporality.

Different Aspects of Temporality of the Cardiac Electric Field.

The analysis of manifestations of temporal properties in the presentation of the cardiac electric field delivers two phenomena. They are discriminated by their effects - either within one single cycle of excitation - or on the influencing of heart rhythm over a series of cycles. The methods of their investigation are different and the information of behaviour or generation of the cardiac electric field points into different directions.

Temporal Behaviour of the Cardiac Electric Field Within One Single Cycle.

As excitation spreads over the heart along a defined pathway during each cycle the body surface image of the field follows this pattern of changes in the configuration of the electrical source. That leads to a series of maps each of them representing one instantaneous state of the field. The sequence of maps reflects the way of the excitation front and can be considered as the imaging of the variations of the equivalent dipole in magnitude and localization. The number of maps necessary for sufficient resolution of the process is relatively high. While a series of maps produces difficulties of interpretation in consequence of the large amount of data presented, some attempts have been made to filter relevant information and to compress it into shorter representations. On the other hand, theoretical investigations have been performed for modelling the course of excitation in the heart and for constructing its reflection in body surface maps (e.g. 6 in 7).

The time course of the cardiac electric field within one cycle shows 3 parts (2). Each part contains a series of body surface maps. The first one reflects the atrial excitation and is characterized by one maximum and one minimum. Both have small intensities of less than 0.1 mV and are situated on the front of the chest surface. The negativity begins its course at the right shoulder going downward and to the left, pushing the maximum over the left side to the back of the thorax. The second part expresses the activation of the ventricles. Now larger extrema of clearly more than 1 mV are measured. The negativity starts at the back, goes over the right shoulder with a quick increase in magnitude, shifts to the precordial region and vanishes there after about 100 ms. The positivity appears precordially and migrates leftward to the back of the thorax. The third part shows moderate potentials up to 1 mV. Its negativity appears sometimes within the end of the second phase on the right shoulder. The positivity lies on the lower parts of the front and at the back of the thorax. That part coincides with the recovery

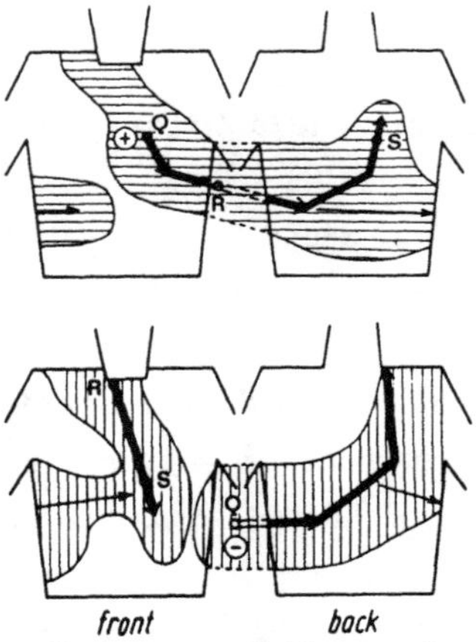

Fig. 1. Hodogram of the maximum (top) and minimum (below) of the cardiac electric field during QRS.

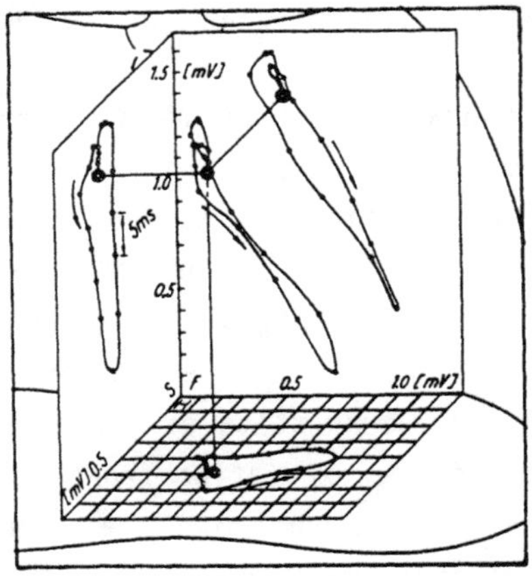

Fig. 2. The vectorcardiogram of the QRS complex in the space of the thorax with its frontal (F), spagittal (S) and horizontal (H) projections.

period of the ventricles. This course is typical for normal persons and can be found even in statistical investigations (8). Physiological influences of respiration concern mostly the size of potentials measured (9). Heart rate changes, e.g. after physical effort, induce a higher velocity in the movement of the field without strong changes of the field pattern (10).

Changes in intensities and localization describe the time course of the cardiac electric field over one cycle representing it as a continuous phenomenon. For the marking of these processes, pathways and velocities have to be stated. Pathways follow the spatial ways of description as shown in the body surface maps series. For demonstrating the whole cycle in only a few concentrated maps the "hodogram" can be proposed, as depicted in Fig. 1. The velocities usually are perceptible in the vectorcardiogram. This reflects the changes of magnitude and spatial orientation of the equivalent vector of the source and is received from leads positioned in the body surface field (s. Fig. 2). The calibration of time within the vectorloops marks equal time distances and thus gives the velocity in the reciprocal of the length of single segments of the course. It becomes clear that the activation of the atria is a slow process, the

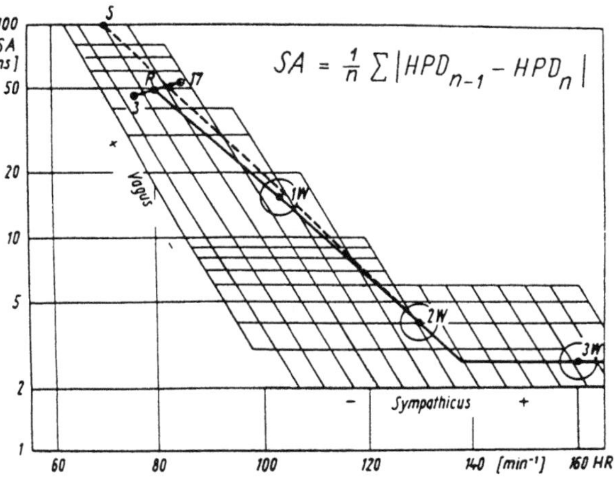

Fig. 3. Rhythm diagram of heart rate vs. sinus arrhythmia under different physical load (1, 2 and 3 W/kg b.w.) in sportsmen (S) and nontrained persons (R). Diurnal shifts of resting values between 3 and 17 hours in R.

beginning and finishing of ventricular activation has a somewhat higher velocity, whereas the changes in the middle part of the ventricular activation reach the highest values of velocity up to about 25 mV/s. These refer mainly to changes in magnitude.

The simplest record of velocities of changes in the field is given by the electrocardiogram. It records the potential changes at one spot or in the difference between two spots on the body surface as a function of time. Here the large QRS-complex expresses clearly the highest velocities during the ventricular activation. In this simple record however, a discrimination between changes in potential or localization is impossible.

Temporal Properties of the Repeated Origin of the Cardiac Electric Field.

The repetition of the triggering of the excitatory cycle generates a regular series of occurrence of the field. This series can be marked by counting the number of cycles per minute, called heart rate. From observations of heart rate during periods of about 1 min, it became visible that the rhythm of the activation of the heart shows a variability instead of behaving constant. The distances between single cardiac cycles, measured as heart period duration, can be utilized for the evaluation of the variability. Different measures are in application. The easiest approach is the interpretation of the variability as a statistical phenomenon. Under such conditions the statistical standard deviation is used. This measure however, is not able to discover shifts of the mean values, which may falsify the results. Therefore, an estimate of the mean of the absolute differences in heart period durations has been applied as applicable for steady conditions as well as for transitory phases, e.g. during onset or offset of physical effort. In investigations of load reactions, training conditions and other physiological circumstances, this measure was called sinus arrhythmia (11). It reflects the total intensity of heart rate variability and has been proven to be extremely dependent on the presence of the cardiac vagal tone (12, a.o). Heart rate on the other hand, underlies to a large part the control of the accelerating sympathetic cardiac activity. Therefore, the correlation of sinus arrhythmia versus heart rate may serve as a combined visualization of the cooperation of both cardiac nerves. Examples of such investigations are demonstrated in Fig. 3. It illuminates the diverse answers of increasing heart rate and decreasing variability under growing physical load and the different intensities of reaction in trained and untrained persons. Even pathological conditions as hypertrophy or mitral stenosis have been investigated and resulted in low resting sinus arrhythmias of only about 10 ms without large changes of heart rate.

Components of the variability of heart rate can be detected in continuous records of heart rate, the cardiotachogram. It makes visible superimposed oscillations of different frequency in the variations of heart rate. The modern statistical method of the spectral analysis by fast Fourier transformation allows the separation of the components of heart rate variability. It results in the spectral density diagram which gives the absolute or relative power of the components of the variability in relation to their frequency. Several such analyses have proven the existence of three main components. Each component is defined by the frequency of its

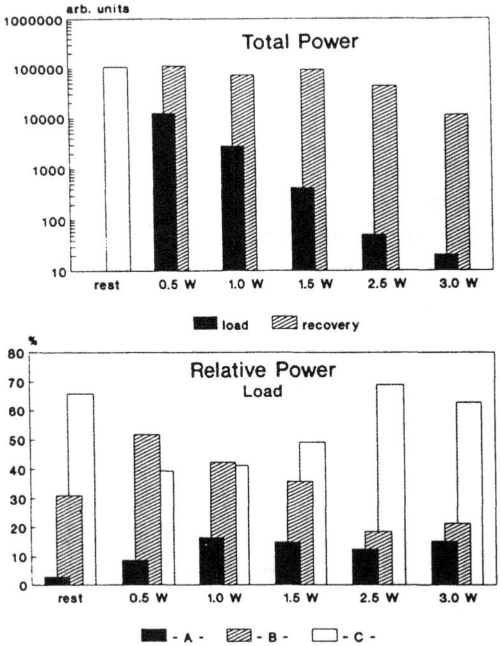

Fig. 4. Spectral density of heart rate variability under physical load, top: total power under load and after 5 min recovery; below: relative power under load in the regions A, B, and C.

maximum spectral density or the frequency range of its appearance and by the amplitude of its maximum expressing the amount of relative power or absolute density within the partition in the whole variability. By several investigations the causing mechanisms for the three components could be determined.

The shape of the spectral density curve gives three maxima (13). The region of the first component is fixed by the range between 0.003 to 0.05 Hz and is labelled as region A. Its maximal power lies about 0.02 Hz. The second component encloses activities within the frequency band of 0.05 to 0.15 Hz and a maximum at 0.1 to 0.12 Hz, called region B. The third component contains the frequencies of 0.15 - 0.50 Hz with its maximum at about 0.2 Hz, called region C. These regions are defined by several investigators in similar values.

Physiological investigations have revealed connections between the three frequency ranges and cardiovascular functions. Region A is closely related to peripheral vascular reactions mainly referring to thermoregulatory activities and encloses sympathetic and parasympathetic participations (13). Region B reflects influences of blood pressure control mechanisms transmitted preponderantly by a parasympathetic drive. Region C is ascribed to respiratory influences by the way of a parasympathetic transmission and reflects the part of the classical "respiratory arrhythmia".

The application of this connection between the generation of the cardiac electric field and cardiovascular regulatory mechanisms opens new ways of investigating load reactions. As an example, a bicycle ergometer test with different loads of 0.5 to 3.0 W/kg body weight in 16 young (age = 25 years) sportsmen are presented in Fig. 4. The load was exerted for 5 min with consecutive 5 min recovery before the next step of load. The total amount of variability corresponding to sinus arrhythmia in Fig. 3 is given as total power. Load reduces it more as the load intensity increases. Recovery restitutes total power practically to the resting value. For relative power, the slow resting component of region A increases its portion until 1W and remains stable for higher load intensities. Component B increases its portion under light load and decreases it in heavy load. The component C mirrors B and decreases under light and recovers in heavy load. All these changes occur under vigorous decrease of total power, which declines by nearly 4 orders of magnitude. 5 min of recovery only under light load restitute the relation between the components, whereas under moderate load, the total power recovers completely without a restitution of the relative power distribution.

The effects of load become more distinct in the spectral density diagram of Fig. 5. The decrease of total power is visible in the changes of calibration. The well expressed relative increase of component C with growing load intensity is accompanied by a shift of its peak frequency up to 0.5 Hz. This reflects the increase of respiratory frequency with growing load, accentuating the coordination of respiration to component C even under heavy load. This effect is demonstrable only by the way of the frequency analysis as total power and sinus arrhythmia disappear nearly completely under

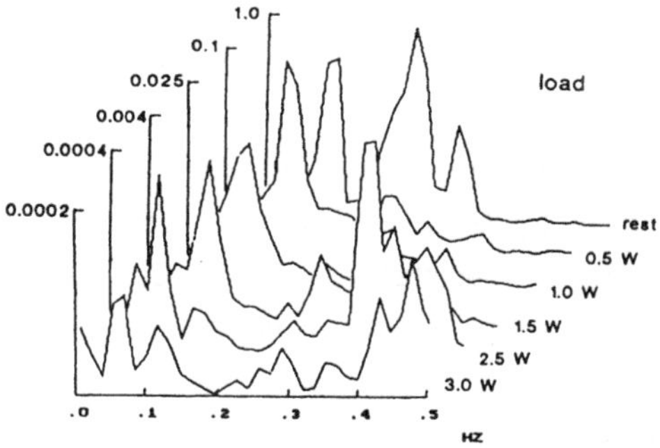

Fig. 5. Spectral density diagrams for heart rate variability in rest and under different load (as indicated). Component A is filtered. Component C shifts under heavy load to 0.5 Hz.

high load. Interpretations of a different participation of regulatory mechanisms in this control of heart rate are obvious.

Different Horizons of Temporality of the Cardiac Electric Field.

Utilizing various methods of the analysis of temporal behaviour in the investigation of the different properties of the cardiac electric field manifests temporal structures and substructures besides the topographic information presented by the spatiality. They are connected to the excitatory process in the heart and to influences on the control of the activation cycle. The most important diversity consists in the various temporal horizons of different order. They require diverse methods of analysis and open insights into various biological processes.

The investigation of the cardiac cycle in the electric field connects the spatial aspects of potential distribution on the body surface - expressing the partition of the source into resting an active myocardium - with the temporal aspects of shifts in localization as well as of changes in magnitude and their velocities. Thus spatiality and temporality were combined and give complete description of the events of one cardiac cycle represented in the cardiac electric field. The temporal order of this representation comprises about 1 s.

The direct temporal analysis of the series of the origin of the cardiac electric field by investigating heart rate and its variability offers another temporal aspect. Thereby one may receive information about mechanisms involved in the control of the cardiac electric field. The temporal horizon for the application of this method is defined in the time interval of about 10 s. The information available by this procedure of temporal analysis concerns dynamics of the complex regulation of the generation of the cardiac electric field in the cooperation of vagus and sympathicus and allows a qualitative estimate of this cooperation. Attempts for quantitative evaluations, e.g. by a statistical treatment of the correlation curve between both parameters is not suitable for receiving unambiguous results.

The application of dissecting the components of the heart rate variability by fast fourier transformation widens quantitatively the evidences of influences upon the temporal determination of the cardiac electric field. With this method the activity of three categories of control mechanisms, namely respiratory, blood pressure and peripheral vasomotoric related effects are separated. Sympathetic and parasympathetic participation in these can be identified. The temporal horizon of this analysis comprises 1 min. Within this time interval, the separated activities became effective. The importance of this method lies in the opportunity for long time investigations of the organization of the generation of the cardiac electric field under influences of adaptory events to load and other physiological reactions of the cardiovascular system and the body.

The application of the three different approaches to the analysis of temporal phenomena in the cardiac electric field completes the opportunities of gaining information about the cardiac activity. Thus it represents a method comparable to the data processing aided procedures of the structural analyses of the cardiac field resulting in equipotential maps, integral maps, departure maps and isochrone contour maps (14). The investigations of spatiality as well as the described ones of temporality of the cardiac electric signals thus open many aspects of diagnostic estimations for the heart itself and the regulation of its performance.

References

1. Waller AD. Philosoph Trans 1989;180:169-176.
2. Taccardi B. Circ Res 1963;12:341-352.
3. Einthoven W et al. Pflug Arch Ges Physiol 1913;150:275-315.
4. Scher AM et al. Circ Res 1953;1:539-547.
5. Durrer D et al. Circulation 1970;41:899-912.
6. Aoki M et al. in (7), 1986;274-282.
7. van Dam TH, van Oosterom A (eds.) Electrocardiographic body surface mapping. 1986; Nijhoff, Dordrecht.
8. Green LS et al. in (7), 1986;9-18.
9. Schubert E et al. Japan Heart J 1982;23(Suppl. 1):388-390.
10. Schubert E et al. In: Models and Measurement of the Cardiac Electric Field (ed.) E Schubert. Plenum, New York, 1982;121-130.
11. Schubert E. Bull Acad Med Bel 1981;13:195-206.
12. Warner HR et al. J Appl Physiol 1962;17:349-355.
13. Akselrod S et al. Science 1981;213:220-222 and Am J Phsyiol 1985;249:H867-H875.
14. Taccardi B. in (7), 1986;3-8.

Computer Modelling of Cardiac Arrhythmias

M. Malik

Department of Cardiological Sciences, St. George's Hospital Medical School, London SW17 ORE, ENGLAND

Introduction

The following text is a review of mathematical models and computer simulation projects aimed at reproducing cardiac arrhythmogenic processes, operation of artificial pacemakers, and electrophysiological mechanisms related to high energy defibrillation. The aim of the text is to present key achievements in this area of basic electrophysiological research and to demonstrate future possibilities of cardiac computer models.

Simulation of Arrhythmias

The very first attempts at artificial reproduction of electrocardiographic records in the late 1920's also considered cardiac rhythm disturbances [1,2]. In this first model, two physical components represented the atria and ventricles. By varying the obstacles between them, atrioventricular block was reproduced; additional activation of either of the components modelled premature beats, etc. This simple principle of arrhythmia simulation has been employed many times in more recent studies. For instance, by 1940 coupled relaxation oscillators were used to simulate ventricular rhythm disturbances resulting from the interaction of two desynchronized pacemakers with different intrinsic rates [3].

Hence, the resemblance between natural and disturbed cardiac rhythms and the behaviour of combined physical oscillators can be used to demonstrate patterns similar to those produced by arrhythmias. On the other hand, this resemblance is too coarse to permit modelling aimed at investigating the details of arrhythmogenic processes. From the practical point of view, these physical 'demonstrative' models have mainly attracted an educational use [4,5]. A similar principle has also been applied to the development of electronic cardiac rhythm simulators used for preclinical evaluation of complex pacemakers which were later converted into personal computer supported systems [6]. However, many other models have been used for studies oriented specifically to the investigation of the cause and nature of arrhythmias.

Computer Modelling of Reentrant Waveforms

The theoretical examination of arrhythmogenic processes started with investigations of anatomically determined reentrant excitations which were the first arrhythmogenic mechanism to be understood in sufficient detail. Reentrant excitation based on a undirectional block in a circuit around an inexcitable region is not only a feature of cardiac tissue. Before it was shown in strips of heart muscle by Mines and Garrey [7-9], it was observed in jellyfish [10]. More recently, the same phenomenon was demonstrated in an earthworm nerve [11]. Anatomically enabled reentry is therefore a general mechanism and it can be artificially reproduced and its features investigated by computer modelling.

Surprisingly, the first models of reentry and even of reentry degenerating into fibrillation were not mathematical or computational but physical. It has been known for many years that the electrochemical processes on the surface of an iron wire placed in nitric acid resemble the action potential of a nerve [12]. This rather remarkable observation has been exploited in many biophysical studies and several of them were related to reentry and fibrillation. The surface of an iron ball was used to demonstrate how fibrillation can be initiated by rapid stimulation or by increasing the excitability of the modelled tissue [13]. Remarkable stimulation of reentry has also been accomplished with a mesh of iron wires which showed not only circular activation waves but also spiral waves rotating around one end of the excitation region [14,15].

This first physical model based on observation of reentry mechanisms without a central inexcitable area (only later called vortex reentry or leading circle reentry) was followed by computer simulations of excitable media [16,17] and theoretical mathematical considerations [18,19] all of which suggested that the central anatomical obstacle was not a crucial requirement for the existence of reentry. Similar rotors were then computed in models of homogeneous cardiac tissues based on the Hodgkin-Huxley type of equations [20,21]. The vortex reentry phenomenon was only observed experimentally in sheets of rabbit atrium [22-24] approximately 10 years after models first predicted its existence.

The model of Moe and co-workers [17] is probably the most well known among the models which predicted vortex reentry. It was also a precursor of many later modelling studies and its ideas represent one pole of the approaches to the simulation of arrhythmia. Myocardial tissue was represented by a two-dimensional sheet made of almost one thousand units ordered in a hexagonal mesh which enabled circular excitation waveforms to be more closely simulated than in a rectangular network. The behaviour of the 'cellular' units was not described by any equations derived from physical theories and measured parameters, but by sets of logical rules. Each unit had an individually programmable refractory period divided into an initial absolute refractory period and a terminal relative refractory period, the duration of which was the same for all units. After receiving an excitation pulse, a resting unit became excited, sent the activation pulse to its neighbouring 'cells', and remained in an inexcitable state for the absolute refractory period after which its excitability was restored. During the relative refractory period, the unit was excitable but accepted the excitation pulse from the neighbourhood with a longer delay in order to model the slower propagation speed of a premature activation wavefront. The refractory period of each unit was also proportional to the square root of the length of the preceding 'activation' cycle. When introducing extensive zones of conduction blocks into the modelled sheet of tissue, the model was capable of reproducing circus movement reentry. However, when introducing a random distribution of refractory periods of individual units (more precisely a random distribution of constants expressing the dependence of refractory periods on the cycle length), the model exhibited patterns of vortex reentry which degenerated into an almost chaotic fibrillation-like behaviour.

The theoretical work related to this model [25] suggested the rotor hypothesis of fibrillation, i.e. that the fibrillation is maintained by a large number of small and rather regular excitation waves, the existence of which is made possible by the dynamic heterogeneity of tissue excitability and conduction velocity. Such a hypothesis is difficult to prove by biological observations but has been extensively examined theoretically by mathematical equation models which describe the propagation of excitation in both homogeneous and non-homogeneous media, including 3-dimensional models [15,26]. Recent publications in this field reported classification of the timing and size conditions under which an extra stimulus will initiate rotor tachycardia. Some of these theoretical predictions were verified in direct experiments with dog hearts [27].

The basic ideas of the initial Moe model have been repeatedly explored in different settings [28]. More powerful computers have allowed an increase in both the number of cellular elements and the duration of simulated experiments. An example of a more recent utilization of the original idea is the model developed at the Massachusetts Institute of Technology [29], which used a rectangular network of a programmable number (a few thousand in practical experiments) of cellular elements shaped into a cylindrical form. Instead of the uniformly random distribution of refractory periods, this model introduced a possibly more realistic normal distribution and was used in experiments examining the relations between the initiation of reentry and fibrillation and the values of the programmable parameters of individual cells [30]. An extension of the basic Moe's concept was also incorporated into other models [31,32] which reflected realistic geometry of the heart. These models were used to examine tachycardia initiation patterns and tachycardia and fibrillation thresholds (lowest current magnitude of a train of pulses which induces tachycardia or fibrillation) in ischemic hearts [31,33] and energy thresholds for ventricular fibrillation [34].

Advances in computing possibilities have also enhanced the ability to model the relative refractory period by incorporating a clinically relevant equation which governs the dependency of the conduction velocity on the length of the preceding cycle [31]. More accurate relationship between the characteristics of elements representing normal and infarcted tissue have been introduced in this way.

Simulation of Interactions between Myocytes

The logical rules governing the behaviour of individual cells in the discrete model of Moe and co-workers reflected [17], in a simplified but satisfactory way, both the cycle length related duration of refractory periods and the slow propagation of premature excitations. However, the interactions between neighbouring cells were limited to the transmission of excitation pulses. During the refractory periods, the neighbouring cells did not interact. Hence, the electrotonic interactions were not properly simulated and the omission of electrotonic interaction can jeopardize the value of arrhythmia models as became obvious in studies published much later [35].

Electrotonic interactions were incorporated into reentry models from another group. This group presented a dual approach to the cellular logical models. The initial model was that of van Capelle and Durrer [36]. They represented a two dimensional sheet of the modelled tissue by a rectangular array of elements which were interconnected by electrical resistances. The behaviour of each element was described by two differential equations corresponding to a simplified version of the classical Hodgkin-Huxley membrane kinetics. The elements then behaved as individual voltage sources and their action potentials only remotely resembled the action potentials of myocytes. Thus the reproduction of individual elements was less accurate than in the earlier model of Moe and co-workers [17]. On the other hand, the description of separate elements as voltage sources enabled the equations modelling individual cells to be combined with known physics based equations describing the electric flow along resistors connecting the modelled cells [20]. As a result, this model strongly reproduced electrotonic interaction between simplified cellular models. The modelled action potentials exhibited typical electrotonic phenomena such as prolongation of the action potential of the cell excited first and shortening of the action potential of the cell excited last.

These ideas of the model were applied to study electrotonic interaction between pacemaker and nonpacemaker cellular elements and between pacemaker cells with different periods and circus movement excitation in a square array of 225 elements. The experiments with tight coupling suggested that a nonpacemaker cell could suppress the automaticity of the pacemaker; the sheet model confirmed the existence of a vortex reentry without a fixed central barrier and even without any special pre-defined distribution of refractory periods of individual elements. The electrotonic interaction which was reproduced enabled the model to simulate situations where the elements in the center of a reentrant vortex were kept in a semi-excited nonpropagating state. Other experiments with the initial model examined the influence of pacemaker cells suppressed by tight coupling on the tissue model with a pre-set inexcitable (ischemic) region [37]. These examinations led to a suggestion that electrotonic radiation of subthreshold depolarizations from the inexcitable region could trigger the automaticity of a pacemaker cell which was suppressed by tight coupling. This was proposed as a mechanism of ventricular ectopic activity after acute ischemia.

From what has been said, the direction in which these models have further been elaborated is obvious. Instead of using simplistic versions of the Hodgkin-Huxley membrane kinetics, the equations describing the action potentials of individual cellular elements have been taken from ionic cellular models [38]. The theoretical descriptions of electrotonic interaction and of excitation propagation in anisotropic tissue have also been further developed [39] and permit much more realistic models to be constructed [40].

These models have been repeatedly used in studies examining the propagation of excitation under different circumstances. Their arrhythmia related applications for instance predicted marked differences (which were confirmed experimentally in canine tissue preparations) in the possibilities of reentrant circuit formation for barrier and excitation propagation along and across myocardial fibre bundles [41]. Another study suggested an explanation of the arrhythmogeneity of loose cellular coupling when showing that cellular uncoupling can unmask intrinsic differences in action potential duration and refractory periods [42].

In fact, the recent mathematical theory of rotors which were mentioned in the previous section, also utilizes a simple mathematical description of electrotonic interaction. Theoretical work based on similar equations to those used for the description of excitation transmission, in excitable media (such as in the chemical reactions) has also been applied to the cardiac muscle [423,44]. The authors of this work proposed an interesting concept that the excitation wavefront radiating through a narrow path (such as a cable of myocytes surrounded by post-infarction fibrosis) could be blocked, not by tissue refractoriness but because of the limited energy in the wavefront when suddenly reaching a wide area of excitable tissue. A narrow pathway opening into the excitable tissue abruptly at one end and gradually at the other end is functionally unidirectional, giving a pathophysiological basis for a reentrant tachycardia [45].

Mathematical abstraction of the mechanisms of anatomically determined reentry leads to the investigation of uni-dimensional circular paths. The easiest approach to such studies could be based on the so-called continuous cable theory [46] which describes the propagation of excitation and repolarization waves along cellular membranes. However, it has been shown experimentally that the theory of continuous excitation transmission cannot explain some electrical properties of myocardial tissue [47,48]. This observation initiated modelling studies examining the differences between the behaviour of continuous and discontinuous reentrant paths [49]. The simulated discontinuities of reentrant paths, i.e. the models of coupling of cells creating the reentry circuit, enabled the influence of coupling on the tachycardia mechanism to be investigated [50,51]. The results of the study were in good agreement with the clinical experience and showed that increased resistances (decreased coupling) between the cells of the path enhanced the vulnerability to reentry measured as the duration of the window during which an extrastimulus initiated the tachycardia.

Finally, an attempt to combine a detailed model of electrotonic interactions with an elaborated version of the discrete logical model of individual cells has also been reported [52]. This model introduced the possibility of independently investigating excitation transmission and electrotonic interaction during the repolarization phase. It proposed a biologically unknown mechanism of depolarization reflection when the electric flow between coupled depolarized and repolarized cells resulted in a new excitation wave. The model then showed an antiarrhythmic effect of electrotonic interaction and anti- and proarrhythmic effects due to the shapes of action potentials [53,54] but whether this simulated arrhythmia mechanism is genuine or artifactual remains questionable.

Models of Junctional Tachycardias

Initial computational models of junctional tachycardias were simple and merely investigated interactions between anatomically determined portions of reentrant circuits with markedly different properties. The elementary units of these models did not represent individual cells or micro-regions of myocardium but whole anatomically and functionally determined areas of the heart.

The first of these macro-conduction models represented the complete heart by 11 elements modelling sinus node, atria, atrioventricular node, His bundle, three main bundle branches, left anterior and posterior and right ventricular regions, and finally a bypass tract interconnecting the atrial element with one of the ventricular elements [55]. The fact that the elements of the model represented large macroscopic regions meant that their behaviour could not be described by action potentials. Similarly, interactions between connected elements other than detailed intracellular electrotonic interaction were simulated. For these reasons, a description by logical rules was appropriate and three state functioning was introduced: when receiving an activation pulse, each element waited for a certain delay after which the pulse was radiated to neighbouring elements; the element then remained inexcitable for the duration of its refractory period; after the end of the refractory period, the initial excitable state was restored.

The elementary logic of individual elements has often been used in recent studies. However, the lack of a properly simulated relative refractory period required additional ad-hoc rules describing the dependence of conduction delays and refractory periods on the duration of the very last cycle and/or on the overall excitation rate of several recent cycles. These dependencies were not simulated but introduced as black-box features of model elements. When using appropriate descriptions of atrioventricular nodal decremental conduction, the simulation experiments exhibited proper atrioventricular phase shifts including realistic patterns of Wenckebach periods [55].

In order to simulate more complex conduction patterns related, for instance, to the initiation of supraventricular tachycardias by premature ectopics, these models demanded more elaborate descriptions of cycle and rate dependencies, especially of those related to the atrioventricular nodal function. Therefore, the development of the macro-conduction models utilized mathematical descriptions of the atrioventricular nodal function which it partly simulated [56-58]. The discussion of mathematical descriptions of nodal function is beyond the possibilities of this review but many different formulae expressing nodal properties have been used in different models. The possibility of using an experimental curve based on measured data instead of an artificial mathematical formula has also been reported [59].

The application of models based on a fixed structure of individual macro-elements is naturally limited. Studies oriented differently required that different conduction structures were examined. This led to the development of 'meccanno' models which permitted different cardiac images to be assembled and simulated [60,61]. The clinically relevant applications of these models have mainly been related to the simulation of pacemaker performance which will be mentioned later. However, one simulation study [62] has used a meccanno model in order to simulate a series of heart images and to disprove a hypothesis that stable cycle alternation of an atrioventricular reentrant tachycardia could be caused by compensating conduction times on a single intra-nodal pathway, i.e. by the mechanism: short cycle --> short time for nodal recovery --> slow nodal conduction --> long cycle --> long time for nodal recovery --> fast nodal conduction --> short cycle [63]. This modelling study assumed that the atrioventricular node is not composed of a single element exhibiting decremental conduction but of a series of such elements, and showed that the mutual interactions between these elements, which were also theoretically examined [64], would cause the alternations of the tachycardia to diminish and finally peter out.

Supraventricular tachycardias have also been simulated using much more detailed models incorporating precise cardiac geometry [65,66]. These models are capable of reproducing cardiac electric fields and surface electrocardiograms. Apart from this electrocardiographic precision, they do not offer substantial advantages when compared with much simpler building block models.

Mathematical Models of Arrhythmogenesis

The mathematical techniques used in the investigation of arrhythmias and in particular of myocardial fibrillation have recently included novel theoretical methods of nonlinear dynamics. Theories of the fractals [67,68] and of the chaotic behaviour of nonlinear systems [69,70] have been applied to investigate the geometrical and topological structure of cardiac tissues and the electrical manifestation of complex arrhythmias.

These theoretical approaches are not proper simulation models and we mention them briefly only because some of the relevant studies were related to and combined with true modelling. The results of such investigations showed that the behaviour of both cardiac cells and their mathematical models is indeed nonlinear. Very small changes of parameters such as ionic concentrations and ionic flow may produce huge differences in the overall behaviour at the levels of cellular blocks and tissue [71,72]. At the same time this nonlinear mechanism has been produced in both laboratory experiments and in corresponding computer models and it appears that some of the arrhythmic patterns which have seemed completely random and chaotic, can be predicted [73].

Simulation of Defibrillation Processes

The electrophysiological processes related to defibrillation have also been studied by means of mathematical and computer models. In principle, the studies simulating defibrillation are of two kinds.

First, models which incorporate detailed geometry of the heart and other thorax organs and tissues have been used to predict the distribution of the electric field during defibrillatory shocks. Already initial studies have concluded that the commonly used electrode positions are not necessarily the best choice [74]. Later, experimental techniques for the measurement of the electrical fields during defibrillation of an isolated dog heart were developed and their results were used to test mathematical and computer models of tissue conductivity [75]. The recent studies have mainly been performed in order to suggest the optimum electrode sizes and positions for external defibrillation which would concentrate the electric flow almost selectively in the region of the cardiac ventricles [76,77]. Similar studies have been conducted for the combinations of intracardiac electrodes and epicardial or subcutaneous patches [78,79] and some of these results have been verified experimentally in dogs [79].

The second group of models comprises studies investigating the influence of defibrillation (and also simulation) currents on the individual cells of the myocardial tissue [80]. Intrinsic ionic processes and the natural activity of myocytes mean that cardiac muscle is not a simple passive electrical conductor and this fact has to be considered in complex defibrillation models. However, this necessity means that both local cellular processes and global conduction processes determined by the anatomical shape and geometry of the heart must be reproduced. Fortunately, the situation is slightly simpler than in the activation and repolarization models of the complete heart and a mathematical apparatus which should permit the combination of the cellular and anatomical levels has already been proposed [81] and laboratory experiments have verified its appropriateness [82].

Computer Models of Artificial Pacemakers

Several modelling studies have been related to various aspects of the functioning of artificial pacemakers. The majority of these studies examined the logic of pacing algorithms but other models have also been reported, e.g. ionic model based simulations of the electrode-tissue interfaces [83].

The algorithms and operational modes of external and implantable pacemakers have predominantly simulated using the building block models which only recognize resting and excited states of individual heart elements and which we have already mentioned in the section on models of supraventricular tachycardias. To a large extent, the development of these pacemaker oriented models was encouraged by the need for a better understanding of heart-pacemaker interaction. The elaborate modes and algorithms of recent dual chamber pacemakers may lead to very complex episodes of mutual influence between natural cardiac activity and the programmed pacemaker. A computer model which is able to reconstruct propagation within the heart and the response of the pacemaker offers a flexible and powerful tool for studying this complex interaction. The elaborate modes and algorithms of recent dual chamber pacemakers may lead to very complex episodes of mutual influence between natural cardiac activity and the programmed pacemaker. A computer model which is able to reconstruct propagation within the heart and the repose of the pacemaker offers a flexible and powerful tool for studying this complex interaction. Studies based on building-block heart models can obviously reproduce intracardiac conduction processes only at a very global level which does not allow detailed generation of intracardiac electrocardiograms. Pacemaker sensing is therefore greatly simplified in such models. As each pacemaker is an engineered device, its computer model can be a virtual replica and the pacing operation can be simulated meticulously [59].

From the conceptual aspect of model construction and programming, there are not many important differences between the reports of various models. The heart is being represented by different network structures composed of conduction elements. Some models utilize very complicated strategies to account for cycle length and rate dependencies of conduction delays and refractoriness of individual elements [61,84], others permit the use of experimentally measured data [59] or omit cycle length dependencies completely [85]. The models of pacing algorithms range from very simple systems [84] to large libraries of pacing modes which include complex investigational modes [59].

Studies examining the performance of pacemakers on much more detailed models of the complete heart have also been reported [86] but the computational complexity of such detailed models makes such an approach unnecessarily time consuming. The evaluation of pacemaker actions requires that many heart beats must be reproduced. In such episodes (e.g. episodes of pacemaker mediated tachycardia), the nature of intracardiac conduction is virtually the same during each beat and its very detailed simulation is practically purposeless.

The form of output which can be generated by these models varies more considerably than their internal structure. In principle, each simulation experiment traces the excitation and recovery changes of individual elements of the heart model together with changes of the time counters of the simulated pacemaker. Therefore, virtually every model of this type is able to produce a time table giving precise timings of individual events during the simulation experiment. Some models only produce such a table [87,88] but for more complicated experiments, the detailed tracing can be too extensive and difficult to interpret. Other models therefore produce approximate ECG records, the patterns of which are not generated as with the models of cardiac field but are only composed of waves and complexes taken from pre-defined templates. Different forms of graphical tracing of the pacing algorithm (e.g. simulated marker channel) as well as simulated ladder diagrams can also be produced [89]. Intracardiac electrocardiographic recordings such as the His bundle electrogram have also been simulated [85] in order to make the results of the simulation experiments more understandable.

As we have already pointed out, heart-pacemaker models offer a powerful opportunity for explaining the behaviour of the pacemaker under different circumstances. Accordingly, the practical aim of many of these models has been educational [90,91]. The output of these teaching models is usually screen oriented and some educational systems have been brought to the level of very high fidelity being able to reproduce many stored episodes of pacemaker function and malfunction, as well as permitting an interactive operation when teaching, for instance, pacemaker programming (Keller W, personal communication, 1991).

Fewer modelling studies have been oriented to pacemaker research although the models can predict the behaviour of investigational pacing modes under different circumstances. Dassen and co-workers suggested a system for computer supported pacemaker design based on a combination of a model and an artificial intelligence founded knowledge base [92]. Their idea was to describe the requirements for the pacemaker operation and then to use the knowledge base of pacemaker actions in order to propose the implementation of a new and automatically developed pacemaker and to generate its computer model. Comprehensive testing (on a heart model) of the pacemaker mode created in this way extended the knowledge base of its properties and led to improved implementation and an improved model. The authors suggested that by repeating this development loop, a new pacemaker could be designed very efficiently.

Another heart-pacemaker model has been used in a series of reports simulating the prevention of atrioventricular junctional tachycardias by dual chamber DDD pacing with short atrioventricular delays [93-95]. Based on the simulated rhythm episodes, the authors of these studies suggested that the programming of the DDD pacemaker is critical for success in preventing the tachycardia and that the currently manufactured DDD pacemakers are unlikely to prevent tachycardias which are initiated by more than one atrial premature depolarization. The final study of this series also proposed a new modification of the DDD pacing algorithm which was, when examined by a computer heart model, more successful in tachycardia prevention than the existing pacemakers.

Finally, an attempt to diagnose pacemaker behaviour by its modelling and by comparing the results with real records of paced electrocardiograms should be mentioned [96,97]. This computer system was capable of a comprehensive analysis of a complicated paced electrocardiograms, the results of which were compared with different possibilities of pacemaker function. The results of the system supported the medical explanation of complex electrocardiographic episodes of paced rhythms. The discrepancies between the ECG analysis and the pacemaker model suggested for instance, loss of sensing or of capture or another pacemaker malfunction.

Perspectives

The future possibilities of computer models of cardiac function and particularly of arrhythmia processes are extensive. The limitations imposed by computational demands are diminishing with the increasing power of modern computers. The performance of existing computers has increased more than one hundred fold during the past decade and this applies not only to top performance super-computers but also to the personal work stations and mini-computers which are affordable for cardiological research and clinical institutions.

This does not mean that we will soon be able to model the complete heart at an ionic level. Reproduction of one cardiac cycle with a truly ionic model of the whole heart with an appropriate number of active cells would still take several decades if not centuries, even when using a network of the most powerful super-computers. On the other hand, models which simplify either the cellular behaviour or the geometry and the topological properties of the cardiac tissue have already reached a level at which they can be used clinically.

The increased performance and high fidelity of recent recording techniques makes it possible to produce accurate maps of cardiac electrical activity. However, these detailed recordings, such as the intracardiac maps of arrhythmic regions, may be very difficult to interpret. Nevertheless, patient oriented models which could be used to reproduce the observed episodes of cardiac function should prove to be of immediate use. Comparisons of the real map with the result of modelling studies could aid the diagnosis of the map. Initial attempts in this direction have already been reported using a simple equation model of the dependencies of refractory periods on the previous activation cycle [98]. Extension of model capabilities and especially the development of an interactive approach with real time interpretation of endocardial activation maps could lead to extremely useful systems.

The future clinical potential of ionic models lies in experiments predicting the effects of drugs [99]. Especially when combined with models examining various aspects of the arrhythmogenic conditions [45,71], the drug related ionic simulation studies could lead to important suggestions such as to the development of preparations; the therapeutic effects of which would be triggered by the fast rate during tachycardia.

Similar potential can be foreseen for models of pacemakers. Again, the results of electrophysiological studies can be used in simulation experiments performed specifically for individual patients. The modelled results could then suggest not only programming of complicated pacemakers but possibly also the placement of sensing and pacing electrodes.

Naturally, basic research in the area of computer models of the heart must continue in order to achieve all these clinical advantages. Steadily increasing interest in 'mathematical and computational cardiology' suggests that this area is capable of important achievements in the years to come.

References

1. van der Pol B, van der Mark J. The heartbeat considered as a relaxation oscillation, and an electrical model of the heart. Philos Mag 1928;6:763-775.
2. van der Pol B. On "relaxation-oscillations". Philos Mag 1926;2:978-992.
3. Bethe A. Die biologischen Rhythmus - Phaenomene als selbstaendige bzw. erzwungene Kippvorgange betrachtet. Pfluegers Arch Physiol 1940;244:1-42.

4. Roberge FA, Bhereur P, Nadeau RA. A cardiac pacemaker model. Med Biol Eng 1971;9:3-12.
5. Bhereur P, Roberge FA, Nadeau RA. A simulation unit for cardiac arrhythmias. Med Biol Eng 1971;9:13-21.
6. Hursta WN, Wells RT, Steinhaus BM. Real-time PC-based heart simulator. Abstracts of the AAMI Meeting, 1991; in press (abstr).
7. Mines GR. On dynamic equilibrium in the heart. J Physiol 1913;46:349-383.
8. Mines GR. On circulation excitations on heart muscles and their possible relation to tachycardia and fibrillation. Trans R Soc Can 1914;4:43-53.
9. Garrey WE. Nature of fibrillary contraction in the heart. Am J Physiol 1914;33:397-414.
10. Mayer AG. Rhythmical pulsation in scyphomedusae. Papers Torgugas Lab Carnegie Inst Wash 1908;1:115-131.
11. Arshavskii YI, Berkinblit MB, Kovalev SA, et al. Periodic transformation of rhythm in a nerve fiber with gradually changing properties. Biofizika 1964;9:365-371.
12. Ostwald W. Periodische Erscheinungen bei der Aufloesung des chrom in Saeuren. Zeit Phys Chem 1900;35:33-76,204-256.
13. Smith EE, Guyton AC. An iron heart model for study of cardiac impulse transmission. Physiologist 1961;4:112 (abstract).
14. Nagumo J, Suzuki R, Sato S. Electrochemical Active Network. In: Notes of Professional Group on Nonlinear Theory of IECE, February 26, 1963; (in Japanese; cited in [14]).
15. Winfree AT. The Geometry of Biological Time. 1980; New York: Springer.
16. Farley BG, Clark WA. Activity in networks of neuron-like elements. In: Information theory, 4th London Symposium. (ed.) Cherry C. 1961; London: Butterworth's.
17. Moe GK, Rheinboldt WC, Abildskov JA. A computer model of atrial fibrillation. Am Heart J 1964;67:200-220.
18. Balakhovskii IS. Several modes of excitation movement in ideal excitable tissue. Biophysics 1965;10:1175-1179.

19. Krinskii VI. Spread of excitation in an inhomogeneous medium. Biofizika 1966;11:676-683.

20. Gulko FB, Petrov AA. Mechanism of formation of closed pathways conduction in excitable media. Biofizika 1972;17:261-270.

21. Shcherbunov AI, Kukushkin NI, Saxon ME. Reverberator in a system of interrelated fibers described by the Noble equation. Biofizika 1973;18:519-525.

22. Allesie MA, Bonke FIM, Schopman FJG. Circus movement in rabbit atrial muscle as a mechanism of tachycardia. Circ Res 1973;33:54-62.

23. Allesie MA, Bonke FIM, Schopman FJG. Circus movement in rabbit atrial muscle as a mechanism of tachycardia. II. The role of non-uniform recovery of excitability of the occurrence of unidirectional block, as studied with multiple electrodes. Circ Res 1976;39:168-177.

24. Allesi MA, Bonke FIM, Schopman FJG. Circus movement in rabbit atrial muscle as a mechanism of tachycardia. III. The "leading circle" concept: a new model of circus movement in cardiac tissue without the involvement of an anatomical obstacle. Circ Res 1977; 41:9-18.

25. Moe GK. On the multiple wavelet hypothesis of atrial fibrillation. Arch Int pharmacodyn 1962;140:183-188.

26. Winfree AT. When Time Breaks Down: The Three-Dimensional Dynamics of Electrochemical Waves and Cardiac Arrhythmias. 1987; Princeton: Princeton University Press.

27. Winfree AT. Electrical instability in cardiac muscle: phase singularities and rotors. J Theor Biol 1989;138:353-405.

28. Swenne CA. Computer simulation of compound reentry. In: Computers in Cardiology New York: IEEE 1987;445-448.

29. Smith JM, Cohen RJ. Simple finite-element model accounts for wide range of cardiac dysrhythmias. Proc Natl Acad Sci USA 1984;81:233-237.

30. Ritzenberg AL, Smith JM, Grumbach MP, et al. Precursor to fibrillation in cardiac computer model. In: Computers in Cardiology (ed.) Ripley KL. New York: IEEE 1984; pp. 171-174.

31. Holley L, Uther J. A computer model of ventricular electrical activity and its application to ventricular arrhythmias. Austral Phys Eng Sci Med 1985;8:88-93.
32. Thakor NV, Eisenman LN. Three-dimensional computer model of the heart: fibrillation induced by extrastimulation. Comput Biomed Res 1989;22:532-545.
33. Fishler M, Thakor N. Tachyarrhythmia threshold of ischemic heart in 3-dimensional computer model. In: Proceedings of the Annual International Conference of the IEEE Engineering in Medicine in Biology Society (eds.) Kim Y, Spelman FA. New York: IEEE 1989;94-95.
34. Province R, Fishler M, Thakor N. Defibrillation threshold simulations on three-dimensional computer heart model. In: Proceedings of the Twelfth Annual International Conference on the IEEE Engineering in Medicine and Biology Society. (eds.) Pedersen PC, Onaral B. New York: IEEE 1990;638-639.
35. Malik M, Camm AJ. Computer model of cardiac repolarization processes and of the recovery sequence. Comput Biomed Res 1989;22:160-180.
36. van Capelle FJL, Durrer D. Computer simulation of arrhythmias in a network of coupled excitable elements. Circ Res 1980;47:454-466.
37. Janse MJ, van Capelle FJL. Electrotonic interactions across an inexcitable region as a cause of ectopic activity in acute myocardial ischemia. A study in intact porcine and canine hearts and computer models. Circ Res 1982;50:527-537.
38. Beeler GW, Reuter H. Reconstruction of the action potential of ventricular myocardial fibres. J Physiol 1977;268:177-210.
39. Barr RC, Plonsey R. Propagation of excitation in idealized anisotropic two-dimensional tissue. Biophys J 1984;45:1191-1202.
40. Roberge FA, Vinet A, Victori B. Reconstruction of propagated electrical activity with a two-dimensional model of anisotropic heart muscle. Circ Res 1986;58:461-475.
41. Kadish A, Shinnar M, Moore EN, et al. Interaction of fiber orientation and direction of impulse propagation with anatomic

barriers in anisotropic canine myocardium. Circulation 1988;78:1478-1494.

42. Lesh DL, Pring M, Spear JF. Cellular uncoupling can unmask dispersion of action potential duration in ventricular myocardium. A computer modeling study. Circ Res 1989; 65:1426-1440.
43. Kogan BY, Karplus WJ, Pang AT. Simulation of nonlinear distributed parameter systems on the connection machine. Simulation 1990;55:271-281.
44. Kogan BY, Karplus WJ, Billett BS, et al. The simplified FitzHugh-Nagumo model with slow recovery properties and 2-D wave propagation. UCLA Technical Report CSD-900016, Los Angeles 1990.
45. Kogan BY, Karagueuzian HS, Karplus WJ, et al. Unidirectional conduction block caused by variations in pathway geometry: A new mechanism for reentry. J Am Coll Cardiol 1991;17:386A (abstr).
46. Jack JJB, Noble D, Tsien RW. Electric Current Flow in Excitable Cells. Oxford: Clarendon Press 1975.
47. Spach MS, Miller WT III, Geselowitz DB, et al. The discontinuous nature of propagation in normal canine muscle. Evidence for recurrent discontinuities of intracellular resistance that affect the membrane currents. Circ Res 1981;48:39-54.
48. Spach MS, Kootsey JM, Sloan JD. Active modulation of electrical coupling between cardiac cells of the dog. A mechanism for transient and steady state variations in conduction velocity. Circ Res 1982;51347-362.
49. Rudy Y, Quan W. A model study of the effects of the discrete cellular structure on electrical propagation in cardiac tissue. Circ Res 1987;61:815-823.
50. Rudy Y, Quan W. The role of cellular discontinuities in reentry of cardiac excitation. In: Proceedings of the Annual International Conference of the IEEE Engineering in Medicine and Biology Society (eds.) Harris G, Walker C. New York: IEEE 1988;946-947.
51. Quan W, Rudy Y. Unidirectional block and reentry of cardiac excitation. A model study. Circ Res 1990;66:367-382.

52. Malik M, Camm AJ. Discrete model of myocardial electrotonic interactions. In: Proceedings of the Twelfth Annual International Conference on the IEEE Engineering in Medicine and Biology Society (eds.) Pedersen PC, Onaral B. New York: IEEE 1990;610-611.
53. Malik M, Camm AJ. Computer simulation of myocardial fibrillation using a one-dimensional model of excitation and recovery processes. Cardiovasc Res 1989;23:132-144.
54. Malik M, Camm AJ. Computer simulation of electrotonic interactions during excitation and repolarization of myocardial tissue. Med Biol Eng Comput 1991; in press.
55. Millane RP, Bones PJ, Ikram H, et al. A computer model of cardiac conduction. Austral Phys Eng Sci Med 1980;3:205-209.
56. Heethaar RM, van der Gon DJJ, Meijler FL. Mathematical model of AV conduction in the rat heart. Cardiovasc Res 1973;7:14-18.
57. Ross DL, Dassen WR, Vanagt EJ, et al. Cycle length alternation in circus movement tachycardia using an atrioventricular accessory pathway. A study of the role of the atrioventricular node using a computer model of tachycardia. Circulation 1982; 65:862-868.
58. Dorveaux L, Heddle W, Jones M, et al. Examination of an exponential model of conduction through the human atrioventricular node. PACE 1985;8:646-655.
59. Malik M, Cochrane T, Davies DW, et al. Clinically relevant computer model of cardiac rhythm and pacemaker/heart interaction. Med Biol Eng Comput 1987;25:504-512.
60. Dassen WRM. A mathematical Model to Study Reentrant Cardiac Arrhythmias. Thesis, University of Limburg, Maastricht 1983.
61. Hagen CT, Dassen WRM, Bump TE, et al. Arrhythmia simulation by computer modeling of cardiac conduction. In: Computers in Cardiology (ed.) Ripley KL. New York: IEEE 1983; 217-220.
62. Malik M, Camm AJ. Compensating conduction times as a mechanism of alternating reentry tachycardia: computer modelling experiments. J Electrocardial 1989; 22:73-80.
63. Schamroth L, Sareli P. Compensating conduction times as a mechanism for alternation during reciprocating tachycardia. J

Electrocardiol 1986;19:291-294.

64. Malik M, Davies DW, Cochrane T, et al. A one dimensional model of atrioventricular nodal conduction. Int J Bio-Med Comput 1987;21:13-32.
65. Malik M, Cochrane T, Camm AJ. Computer simulation of the cardiac conduction system. Comput Biomed Res 1983;16:454-468.
66. Wei D, Yamada G, Musha T, et al. Computer simulation of supraventricular tachycardia with the Wolff-Parkinson-White syndrome using three-dimensional heart models. J Electrocardiol 1990;23:261-273.
67. Goldberger AL, West BJ. Fractals in phsyiology and medicine. Yale J Biol Med 1987;60:421-435.
68. Barnsley MF, Massopust P, Strickland H, et al. Fractal modeling of biological structures. Ann New York Acad Sci 1987;504:179-194.
69. Caplan DT, Cohen RJ. Fibrillation vs. random noise: a comparison using dimensionality calculation. In: Proceedings of the Annual International Conference of the IEEE Engineering in Medicine and Biology Society (eds.) Kim Y, Spelman FA. New York: IEEE 1989;92-93.
70. Holden AV, Lab MJ. Chaotic behavior in excitable systems. Ann New York Acad Sci 1990;591:303-315.
71. Chialvo DR, Michaels D, Jalife J. Supernormal excitability as a mechanism of chaotic dynamics of activation in cardiac Purkinje fibers. Circ Res 1990;66:525-545.
72. Chialvo DR, Gilmour RF Jr, Jalife J. Low dimensional chaos in cardiac tissue. Nature 1990;343:653-657.
73. Vinet A, Chialvo DR, Jalife J. Irregular dynamics of excitation in biologic and mathematical models of cardiac cells. Ann New York Acad Sci 1990;601:281-298.
74. Doian AM, Horacek BM, Rautaharju PM. Evaluation of cardiac defibrillation using a computer model of the thorax. Med Instrument 1978;12:53-54.
75. Kothiyal KP, Shanakar B, Fogelson LJ, et al. Three-dimensional computer model of electric fields in internal defibrillation. Proc IEEE 1988;76:720-730.

76. Fahy JB, Kim Y, Ananthaswamy A. Optimal electrode configuration for external cardiac pacing and defibrillation: an inhomogeneous study. IEEE Trans Biomed Eng 1987;BME-34:743-748.

77. Barnett DW, Fahy JB, Wu H-J, et al. Finite element model applications in defibrillation and external cardiac pacing. In: Proceedings of the Annual International Conference of the IEEE Engineering in Medicine and Biology Society (eds.) Harris G, Walker C. New York: IEEE 1988;200-201.

78. Sepulveda NG, Echt DS, Wikswo JP Jr. Finite element models used for the analysis of cardiac defibrillation. In: Proceedings of the Annual International Conference of the IEEE Engineering in Medicine and Biology Society (eds.) Harris G, Walker C. New York: IEEE 1988;198-199.

79. Sepulveda NG, Wikswo JP Jr, Echt DS. Finite element analysis of cardiac difibrillation current distributions. IEEE Trans Biomed Eng 1990; BME-37:354-365.

80. Plonsey R, Barr RC. Inclusion of junction elements in a linear cardiac model through secondary sources: application to difibrillation. Med Biol Eng Comput 1986;24:137-144.

81. Krassowska W, Pilkington TC, Ideker RE. Potential distribution in three-dimensional periodic myocardium - Part I: Solution with two-scale asymptotic analysis. IEEE Trans Biomed Eng 1990; BME-37:252-266.

82. Krassowska W, Frazier DW, Pilkington TC, et al. Potential distribution in three-dimensional periodic myocardium - Part II: Application to extracellular stimulation. IEEE Trans Biomed Eng 1990; BME-37:267-284.

83. Steinhaus BM. Estimation cardiac transmembrane activation and recovery times from unipolar and bipolar extracellular electrograms: A Simulation study. Circ Res 1989;64:449-462.

84. Dube B, LeBlanc AR. Contribution a la modelisation du systeme de conduction cardiaque. University of Montreal Technical Report IGB-86-11-01, Montreal 1986.

85. Ahlfeldt H, Tanaka H, Nygards M-E, et al. Computer simulation of cardiac pacing. PACE 1988;11:174-184.

86. Malik M, Nathan A, Camm AJ. Computer simulation of dual chamber pacemaker algorithms using a realistic heart model. PACE 1985;8:579-588.

87. Dassen W, Brugada P, den Dulk K, et al. A mathematical model to study pacemaker related tachyarrhythmias. In: Computers in Cardiology (ed.) Ripley KL. New York: IEEE 1983;407-410.

88. Dassen WRM, den Dulk K, Gorgels APM, et al. Evaluation of pacemaker performance using computer simulation. PACE 1985;8:795-805.

89. Malik M. Present status of computer models of heart-pacemaker interaction. Cardiovasc World Report 1989;2:4-10.

90. Fukushima M, Inoue M, Fukunami M, et al. Computer-assisted education system for arrhythmia (CAESAR). Comput Biomed Res 1984;17:376-388.

91. Byrd CB, Byrd CL. A computerized system for modeling pacemaker rhythm. J Electrocardiol 1987;20:Suppl. October:28-33.

92. Dassen WRM, van der Steld A, van Braam W, et al. PACTOT: A reprogrammable software pacing system. PACE 1985;8:574-578.

93. Malik M, Davies DW, Camm AJ. Limiting factors in the use of DDD pacemakers to prevent junctional reentry: computer modeling experiments. Clin Prog Pacing Electrophys 1986;4:137-146.

94. Malik M, Davies DW, Camm AJ. Computer modeling of DDD pacemakers for use in prophylaxis of junctional re-entry trachycardia. PACE 1987;10:839-852.

95. Malik M, Davies DW, Camm AJ. Modification of DDD pacing mode to prevent junctional reentry tachycardia: Computer modelling experiments. PACE 1988;11:1465-1478.

96. Malik M, Camm AJ. The pacemaker inverse problem - computer diagnosis of paced electrocardiograms. Comput Biomed Res 1988;21:289-306.

97. Malik M, Camm AJ. Diagnosis of paced electrocardiograms by inverse modeling of pacemaker actions. PACE 1988;11:2093-2100.

98. Stevenson WG, Nademanee K, Weiss JN, et al. Programmed electrical stimulation at potential ventricular reentry circuit sites. Comparison of observations in humans with predictions from computer simulations. Circulation 1989;80:793-806.

99. Shumaker JM, Clark JW, Giles WR, et al. A model of the muscarinic receptor-induced changes in K^+-current and action potentials in the bullfrog atrial cell. Biophys J 1990;57:567-576.

D. HEART IN CHRONIC HYPOXIA

Relationship Between Heart Mass and Hemoglobin/Hematocrit at High Altitude - Function or Fallacy?

M.T. Kopetzky and S. Daum

Departments of Pathology and Physiology, Texas Tech University School of Medicine, Lubbock, TEXAS 79430, and Klinikum rechts der Isar, Technische Universität München, 8000 München 80, GERMANY

Introduction

Individual differences in the response of the heart mass to exposure to altitude have been known for a century (1). Individual differences in the response of hemoglobin or hematocrit levels to exposure to altitude have been known even longer (2). Little attention has been paid to the possible correlation of the two sets of data. However, at the other end of the hemoglobin scale, in anemia, it was found that heart mass is indirectly proportional to hemoglobin levels (3). We decided therefore to re-examine the heart mass/hemoglobin or hematocrit relations in altitude anoxia.

Materials and Methods

Simulated High Altitude

We used first a low-pressure chamber described by one of us (4), and in later experiments an essentially similar design, except that the air was supplied from a room air conditioner. The chambers were evacuated to a differential pressure corresponding to simulated high altitude of 5,000 meters (SHA) above the control site. Temperature was controlled between 24 and 29°C, relative humidity between 50 and 70%; light and dark cycle was 12/12 hours.

Animals

Male white rats of the Wistar strain were used throughout. Animals entered the experiment with body weights around 120 grams and reached terminal weights of around 300 grams.

Protocol of Exposure

Exposure on alternate days, group pair feeding and control of body weights was as described previously (5). The important point is that terminal body weights of SHA groups and control groups (297 ± 7.0 and 301 ± 5.6 grams) were not statistically different.

Hemoglobin Concentrations and Hematocrit Values

The animals were killed, after barbiturate anesthesia, by cutting the abdominal aorta and bleeding. A small quantity of heparin was injected into lower vena cava before cutting the aorta. The blood accumulated in the abdominal cavity was sampled for the determinations. All were done in duplicates at least. Hemoglobin was determined in the first series of experiments as reduced Hb (6) and in the later series as cyanmethemoglobin (7). Packed cell volume was determined by microhematocrit method.

Heart Mass

After removal of the large vessels and the atria, the ventricular part of the hearts was dissected and the masses of the right and left ventricles and of the septa determined according to Beznak (8).

Resistance to Acute Anoxia of the Right Ventricular Muscle

Isolated right ventricles were suspended in oxygenated Krebs-Ringer phosphate with 5 mM glucose and stimulated 12/min with slightly supranormal square wave shocks using field electrodes. Isotonic and in later experiments isometric contractions at optimal preloads were recorded. After achieving stable output the preparation was exposed to anoxia by bubbling nitrogen through the bath for 60 minutes; then it was reoxygenated for a time needed to re-establish a stable output - usually about 45 minutes. The output after reoxygenation was expressed as percentage of the output before

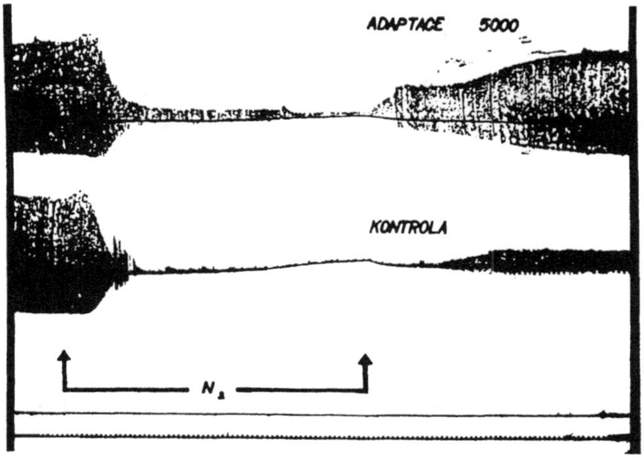

Fig. 1. From the left: above, stabilized output of a right ventricular preparation from an animal acclimatized to SHA of 5,000 m; below, stabilized output of a control preparation; N_2 bracket duration of anoxia; bottom: event signal, time signal. Disappearance of contractile activity during anoxia in both preparations. Reappearance of contractions on reoxygenation; preparation from the acclimatized animal (ADAPTACE 5000) shows much more pronounced recovery than control preparation (KONTROLA). From (10).

anoxia (see Fig. 1).

Results

Hemoglobin and Hematocrit

Hemoglobin concentrations varied in controls from 9.5 to 18.1 g/100 ml, and in SHA animals from subnormal to 28.0 g/100 ml. However; the subnormal values were found in our first few series of experiments in which the reduced hemoglobin method (6) was used. This method yields values not comparable to cyanmethemoglobin method (7) without recalculation. Therefore these data are omitted here and will be the subject of another communication.

Hematocrit values varied of course with hemoglobin concentrations; the control values spread from 26 to 53% and the SHA values from 55 to 82%. Like with hemoglobin, only data from later series of experiments are reported here.

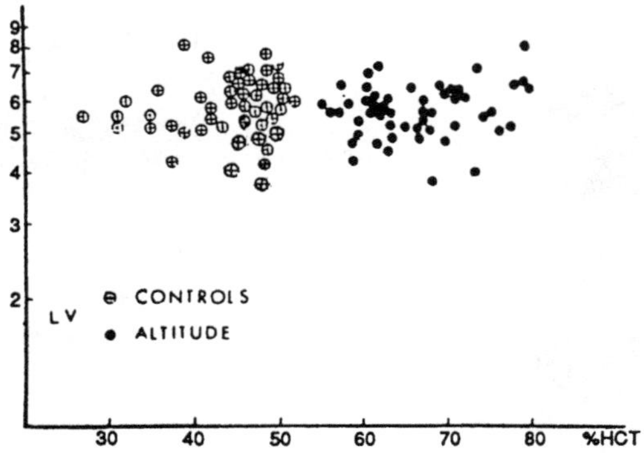

Fig. 2. Abscissa: hematocrit in %; ordinate: absolute left ventricle weights in tenths of gram. Circles are control values, full dots are SHA values.

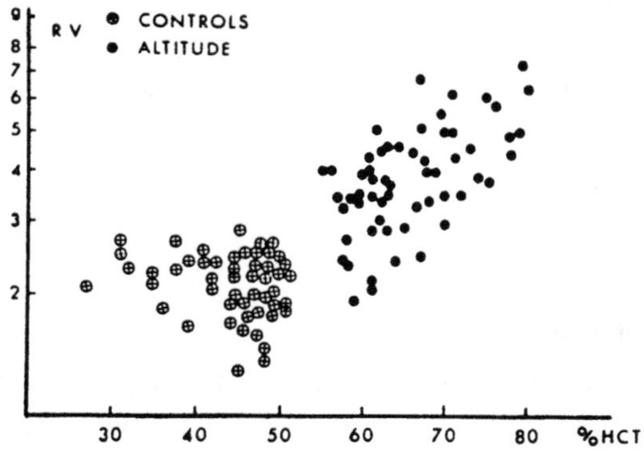

Fig. 3. Abscissa: hematocrit in %; ordinate: absolute right ventricle weights in tenths of gram. Circles are control values, full dots are SHA values.

Heart Mass

Because mean final body weights of control and SHA animals were not significantly different (see Methods), no ratios of heart weight to body weight or other normalization of the figures were attempted.

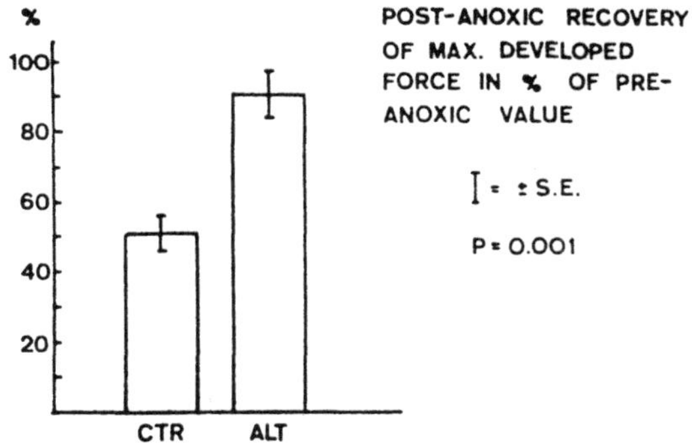

Fig. 4. Comparison of recovery from acute anoxia in control (CTR) and SHA (ALT) preparations of right ventricles.

Absolute wet weights of the left and right ventricles are reported here. The dry weight to wet weight ratios were the same in control and SHA animals.

Left ventricles. The left ventricle masses varied from 0.39 to 0.81 grams in controls, and from 0.39 to 0.82 grams in SHA animals. There is no difference between the two groups.

Right ventricles. The right ventricle masses varied from 0.12. to 0.30 grams in controls, and from 0.19 to 0.72 grams in SHA animals. The difference is significant.

Relationship Between Ventricular Masses and Hematocrit Values

Fig. 2 shows control and SHA left ventricular masses plotted against hematocrit values from the same animal. Obviously the magnitude of hematocrit is not correlated with left ventricular mass either at zero or high altitude.

Fig. 3 shows control and SHA right ventricular mass plotted against hematocrit value from the same animal. There is no correlation of hematocrit with right ventricular mass in controls. However, rising hematocrit in SHA animals is correlated with exponentially rising right ventricular mass.

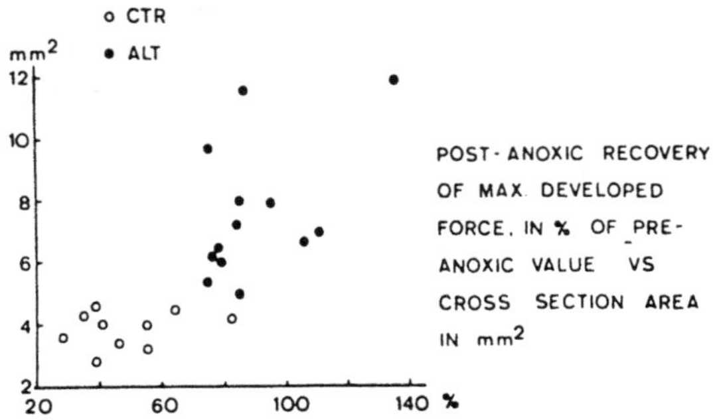

Fig. 5. Abscissa: post-anoxic recovery of maximal developed force in % of pre-anoxic values in controls (CTR - circles) and SHA animals (ALT - full dots); ordinate: square section of the right ventricle preparation in mm square.

Resistance to Acute Anoxia of the Right Ventricular Muscle

Fig. 4 shows that right ventricle preparations from controls recover from 60 minutes of anoxia in vitro to 51% of pre-anoxic maximal developed force. Preparations from SHA animals recover from the same period of anoxia to 90% of preanoxic maximal developed force. The difference is significant.

Discussion

The correlation of right ventricle mass with hematocrit shown in Fig. 3 presents at least one puzzling feature, and that is the rather pronounced variation of the hematocrit and mass figures. There is at the left of the spread of SHA data a group without a significant hypertrophy and with hematocrit values below 60%. Is it possible that these animals were able to compensate by increased ventilation the drop in alveolar partial oxygen pressure? Barometric pressure at altitude 5,000 m above Lubbock, Texas, is 365 mmHg. Therefore, inspired partial pressure of oxygen is 66 mmHg. Assuming that an acclimatized rat can hyperventilate to an alveolar partial pressure of carbon dioxide of 20 mmHg, he would have an alveolar partial pressure of oxygen of 43 mmHg (with R = .85). This is sufficient to keep the mixed venous blood partial pressure of oxygen at a safe level of about 30 mmHg without a significant increase in cardiac output (11). Even so, there is a threshold anoxic stimulus of erythropoiesis - some increase in hemoglobin to 18 - 19 g/100 ml - and of pulmonary vasoconstriction - some, though insignificant, rise in ventricular mass.

At the right end of the spread are animals with a very large ventricle - approaching left ventricle values - and very high hematocrits. Assuming that these animals cannot hyperventilate, how do they use the high hemoglobin? Rat alveolar partial pressure of carbon dioxide reportedly varies from 34 - 42 mmHg (12). Using as before the R = .85 and the more favorable lower figure for carbon dioxide, this translates to an alveolar partial pressure of oxygen of 32 mmHg. From the rat oxyhemoglobin dissociation curve (13) the saturation of hemoglobin with oxygen is then 28%. With the high hemoglobin of this animal (28.0 g/100 ml), his arterial blood contains 10.5 vol % of oxygen. Assuming an arteriovenous difference of 5 vol %, i.e. normal cardiac output, the mixed venous blood partial pressure of oxygen would be about 20 mmHg. The tissues of this animal certainly do suffer from chronic hypoxia.

However, as demonstrated in Fig. 4, the right ventricles of SHA animals possess a resistance to anoxia. More important, the resistance is a function of the square section of the ventricle, i.e. of its size (Fig. 5). In other words, the higher the degree of hypoxia, the bigger the ventricle becomes, and also the more resistant to hypoxia.

It can be argued that ventricles associated with high hematocrits are hypertrophied because of the increase in blood viscosity. However, we submit that this factor might play a much smaller role than generally assumed. First, the effect of hematocrit magnitude on blood viscosity in a circulatory system of such small dimensions and unknown shear rates is not known (14). Second, why are there no hypertrophies in the left ventricles from the same animals (see fig. 2)? Admittedly, we neglect here the possible effects of anatomical changes in pulmonary vasculature in animals exposed to chronic intermittent SHA as described by Ostadal and Widimsky (15). We cannot attempt to analyze the interplay between the magnitudes of hematocrits, changed vasculature and hypoxic constriction.

These considerations led us to propose, in summary, that:
1. The magnitude of hematocrit/hemoglobin levels at SHA is an inverse measure of the success of ventilatory adaptation to SHA. **2.** The degree of alveolar hypoxia goes parallel to the degree of pulmonary vasoconstriction and right ventricular hypertrophy. **3.** Rats exposed to intermittent SHA develop a degree of tissue or organ resistance to hypoxia, commensurate to the degree of tissue hypoxia. **4.** The increased hematocrit in itself is not the cause of right ventricular hypertrophy or only a minor one.

References

1. Strohl JL. Zool Zbl III, 1910;30:1-44.
2. Jourdanet D. In: Influence de la pression de l'air sur la vie de l'homme. Masson, Paris 1895.
3. Chapman DG, Campbell JA. Brit J Nutr 1957;1:117.
4. Kopecky M. J Appl Physiol 1960;15:540.
5. Hughes MJ, Kopetzky MT, Messiha F. J Appl Physiol 1981;51:1607.
6. Heilmeyer L, Mutius V. In: Plöttner K, Klinische Photometrie etc., Fischer, Jena 1944.
7. ICSH, J Clin Path 1978;31:139.
8. Beznak M, Pflügers Arch 1956;262:181.
9. Kopecky M, S Daum. Cs fysiol 1958;7:218.
10. Kopecky M. Acta III Europ de Cor Conv, Rome 1960; Vol 3, p 1025.
11. Fenn WO, Rahn H, Chadwick LE. In: Air Force Technical report 1951;6528:314.
12. Singer RB. In: Handbook of Respiration. (eds.) DS Dittmer, RM Grebe, Saunders, Philadelphia, 1958; p 93.
13. Bartels H. ibidem, p 79.
14. Nichols WW, O'Rourke MF. In: McDonald's Blood Flow in Arteries. Lea & Febiger, Philadelphia 1990; p 12 ff.
15. Ostadal B, Widimsky J et al. Intermittent Hypoxia and Cardiopulmonary System, Academia, Prague, 1985.

Functional Assessment of the Hypertrophic Right Ventricle in the Rat Heart

F. Kolár, B. Ostádal, R, Cihák*, F. Papousek

Institute of Physiology, Czechoslovak Academy of Sciences and
**Postgraduate Medical and Pharmaceutical Institute,*
Prague, CZECHOSLOVAKIA

Introduction

Whereas a number of studies can be found dealing with the function of the left ventricle (LV) in experimental cardiac hypertrophy, considerably less attention has been paid to the right ventricular (RV) performance. This is despite the fact that many pathogenetic factors such as chronic hypoxia (e.g. 1), strenuous exercise (2), excess thyroid hormone or catecholamines (3, 4) are known to induce cardiac hypertrophy with the right ventricular dominance. One of the reasons for this may consist in methodical difficulties associated with the assessment of the RV function, especially in small laboratory animals. Measurements of force/length/velocity relationships on isolated thin preparations of ventricular muscle represent a frequently used approach. Problems arise, however, when the mechanical and pump functions of the whole intact RV are to be assessed under conditions of controlled preload and afterload.

In this study we compared two methods which originally had been developed and routinely employed for evaluation of the LV function. Recently, their modifications were introduced which enable us to study the RV performance. The first method is based on the "in vitro" perfused heart with the RV performing pressure/volume work (5, 6). It represents a modification of the classic (left ventricular) working heart (7) and makes it possible to measure the RV contractile and pump activities under controlled loading conditions. The other method makes it possible to estimate the maximum RV mechanical performance (or reserve) in the "in situ" heart as a response to acute pressure load after a ligation of the pulmonary artery (4). Also this technique is a modification of the procedure used for the measurement of the LV function (8, 9). The rat heart with the RV hypertrophy induced by chronic hypoxic hypoxia was employed as an experimental model.

Animal Model

60-day-old male rats of Wistar strain were acclimatized to intermittent high altitude (IHA) hypoxia simulated in a low pressure chamber for 8 h/day, 5 days/week. Barometric pressure was reduced gradually during 13 exposures up to the level of 306.8 mm Hg which corresponds to an altitude of 7000 m; the total number of exposures was 24-26. Half of the IHA-acclimatized group was employed for the functional study 24 hours after the last hypoxic exposure (hypertrophic group); the remaining animals were transferred into the normobaric conditions equivalent to an altitude of 200 m and kept for a further 5 weeks until employed (regression group). The control group of rats was kept at the altitude of 200 m for the whole investigated period. All animals had free access to tap water and a standard laboratory diet.

Body weight and heart weight paramaters of rats acclimatized to IHA and those examined 5 weeks after the termination of hypoxic exposure are compared with controls in Table 1. The IHA hypoxia induced absolute as well as relative right ventricular enlargement; the complete regression of weight occurred 5 weeks after withdrawal of rats from hypoxic environment.

Right Ventricular Function in Vitro

The function of the RV was assessed using a modified isolated perfused heart with the RV performing pressure/volume work (5, 6). The perfusion scheme of this preparation is illustrated on Fig. 1. The animals were anesthetized with sodium pentobarbital

Table 1. Body and heart weight parameters in the rats of control, hypertrophic and regression groups.

	Control	Hypertrophy	Regression
BW (g)	392 ± 26	337 ± 7*	416 ± 14**
HW (g)	1088 ± 70	1192 ± 39	1227 ± 55
RV (mg)	221 ± 14	309 ± 18*	246 ± 17**
RV/BW	5.70 ± 0.31	9.45 ± 0.59*	5.89 ± 0.27**

BW - body weight, HW - heart weight, RV - right ventricular weight, RV/BW - relative right ventricular weight. Values are means S.E.M., * $P < 0.05$ vs. controls, ** $P < 0.05$ vs. hypertrophy.

(50 $mg.kg^{-1}$), the heart was quickly excised from the thoracic cavity with the lungs and immersed into cold (5°C) saline. The lungs were removed and a cannula was inserted into the aorta and the heart was perfused in a non-recirculating Langendorff mode. The perfusion medium consisted of ($mmol.l^{-1}$): NaCl 118.0, KCl 4.7, $CaCl_2$ 2.5, $MgSO_4$ 1.2, $NaHCO_3$ 25.0, KH_2PO_4 1.2, glucose 7.0, Na pyruvate 2.0 and mannitol 1.1. The medium was saturated with the mixture of 95% O_2 and 5% CO_2 (pH 7.4) and maintained at 37°C. The perfusion pressure was kept constant at 100 cm H_2O.

The remaining fat and connective tissue was then removed from the Langendorff-perfused heart, the inferior caval vein was exposed and the pulmonary artery was cut just before bifurcating into the left and right pulmonary arteries. The superior caval vein and the thoracic veins were ligated. The heart was then attached to the simulated pulmonary circulation consisting of the right atrial filling reservoir and the adjustable tubing system for the control of pulmonary outflow resistance. Another cannula was introduced into the inferior caval vein and connected to the filling reservoir. The level of this vessel could be changed with respect to the heart to allow the right atrial filling pressure (RAFP, preload) variation in the range of -2.5 to 20 cm H_2O. The third cannula was inserted into the pulmonary artery and connected to the tubing system; the outflow resistance could be changed by increasing the pulmonary

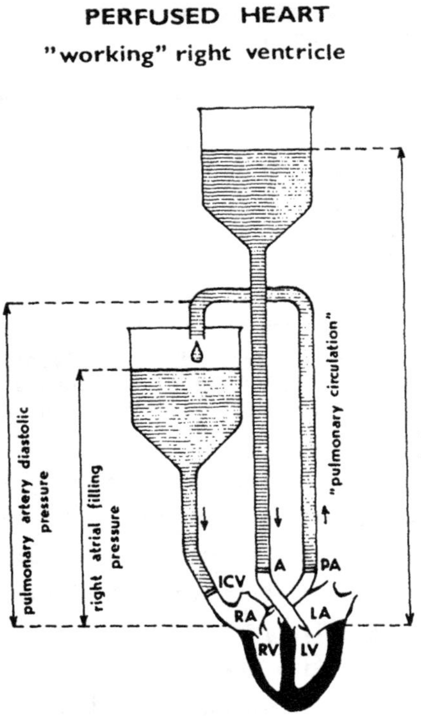

Fig. 1. Schematic representation of the isolated perfused heart preparation with the working right ventricle. RV - right ventricle, LV - left ventricle, RA - right atrium, LA - left atrium, A - aorta, PA - pulmonary artery, ICV - inferior caval vein. Arrows indicate the flow direction.

artery diastolic pressure (PADP, afterload) in 5 cm H_2O increments Starting from 10 cm H_2O.

After completing the cannulations (within 10 min) the heart was electrically stimulated at a constant rate of 310 beats.min^{-1} using a coaxial electrode (Hugo Sachs Elektronik, F.R.G.) attached to the right atrium. The intensity of the stimulating pulses was set at 20% above the threshold value for each individual experiment.

The RV contractile function was evaluated by means of a system with microcomputer which enabled us to measure and analyze the pressure curve. For measurement of the pressure, a 21-gauge needle was inserted into the ventricular cavity transmurally through the apex and connected via a polyethylene cannula to a HP-1280 pressure transducer (Hewlett Packard). After an amplification in a HP-8805B amplifier the electric signal was conveyed to a SAPI-1 microcomputer (Tesla) with an ADC-85C-12 A/D converter (Burr Brown). The amplitude response characteristic of the measuring system was constant in the range of ± 0.5 dB for frequency band up to 30 Hz. The pressure curve was sampled with a frequency of 300 Hz and monitored on the computer screen. The last five cardiac cycles were continuously stored in a circle buffer of the computer memory for the next processing. In two windows of the screen either the pressure and dP/dt or two pressure curves with different time scales could be displayed simultaneously. The selected traces were printed from the screen by the computer graphic printer; in the first experiments a line recorder HP-7702B was used. The stored pressure curve was analyzed in desired time intervals on a command from the computer keyboard. The program evaluated the systolic (RVSP) and diastolic (minimal, RVDP) pressures, developed pressure (DP, calculated as a difference between RVSP and RVDP), and the maximal rate of pressure development [$+(dP/dt)_{max}$]. The index of contractility [+(dP/dt)/P] was calculated as a ratio of $+(dP/dt)_{max}$ to DP. Cardiac output (CO) was measured by a timed collection of pulmonary artery outflow; coronary flow was equal to CO when the filling circuit was clamped.

The RV working preparation was allowed to stabilize for 15 min at the lowest PADP (10 cm H_2O) with closed filling circuit. The RAFP was lowered to -2.5 cm H_2O in order to drain the coronary effluent and unload the ventricle and then gradually increased up to 20 cm H_2O, while the PADP was maintained low (at 10 cm H_2O). As a next step after the volume load test the RAFP was set at 10 cm H_2O and the PADP was stepwise elevated in 5 cm increments until the maximum contractile parameters were recorded and the ventricle failed to reach the given afterload. Functional parameters were computed after reaching steady state at each level of RAFP or PADP (within 3 min). The preparation was stable for the required period of time as confirmed by control experiments.

Effect of IHA Hypoxia

The RV functional performance of rats with IHA-induced hypertrophy and after its regression was compared with that of control animals both during volume load (Fig. 2) and during pressure load (Fig. 3). Hypertrophic hearts achieved higher systolic pressure and $+(dP/dt)_{max}$ at any given preload and afterload and contractile failure occurred at much higher load than in controls. The increase of filling pressure was associated with slight elevation of right ventricular diastolic pressure with no differences among individual groups. On the other hand, the diastolic pressure increased exponentially with increasing afterload in control rats; in hypertrophic ventricles this response was markedly less expressed and shifted to higher afterload. Cardiac output increased approximately 10 fold with the rise of filling pressure from -2.5 to 20 cm H_2O in all three groups. The decline in pumping function with increasing afterload was steeper in controls when compared with hypertrophic ventricles which were able to reach a much higher afterload. Small differences in the cardiac output among individual groups observed at low afterload were abolished when the cardiac output was normalized to body weight (e.g. at RAFP and PADP equal to 10 cm H_2O the cardiac output reached 129 ± 8, 131 ± 5 and 131 ± 6 $ml.min^{-1}.kg^{-1}$ in control, hypertrophic and regression groups, respectively).

The regression of hypertrophy was accompanied by the reversal of ventricular mechanical and pump performance, except for a persisting slight increase of RVSP (or RV developed pressure) at higher load. The maximum values of mechanical parameters achieved at increased afterload are compared in Fig. 4. The index of contractility of hypertrophic ventricles did not differ from controls which indicates that the enhanced ventricular performance is merely the consequence of the increased muscle weight. Hence, a close correlation was found between the ventricular weight and the peak mechanical performance (6). On the contrary, the regression of hypertrophy was associated with small but significant decrease of contractility (by 11%). We suppose that the elevated concentration of collagenous proteins which accompanies the regression of hypertrophy in this particular model (10) may account for this observation.

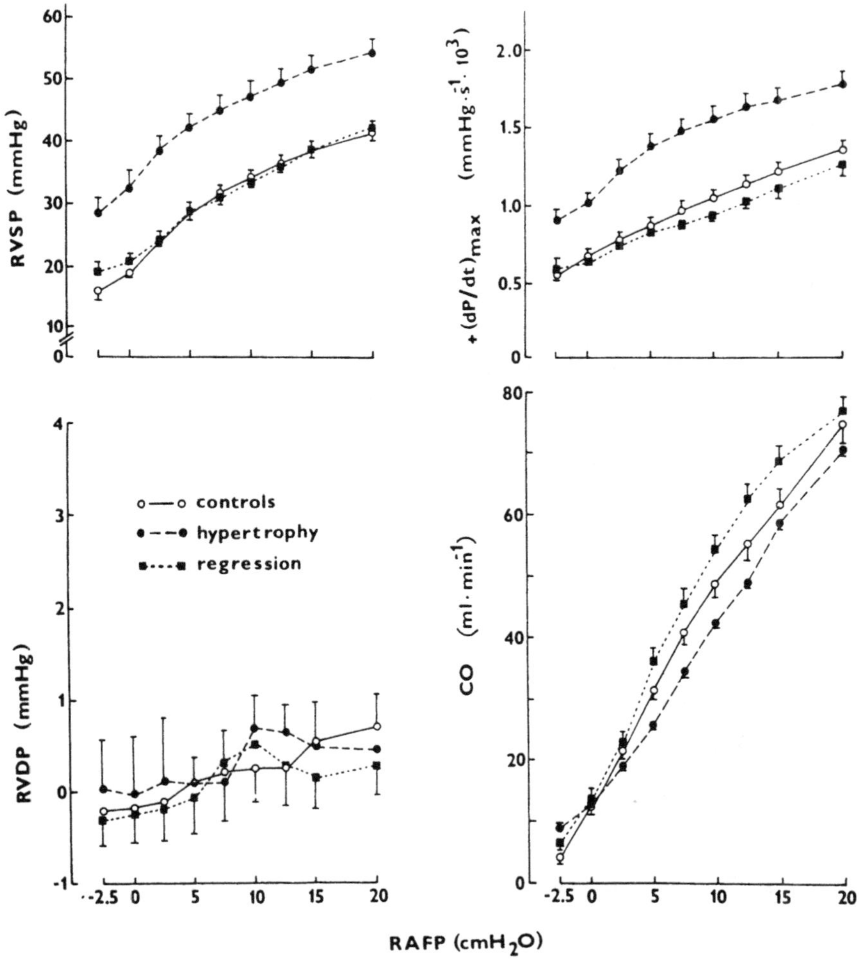

Fig. 2. Changes in the right ventricular functional parameters with an increase with an increase in right atrial filling pressure (RAFP, preload) in hearts of IHA-exposed rats (hypertrophy) and of rats five weeks after the last hypoxic exposure (regression). Pulmonary artery diastolic pressure was set at 10 cm H_2O. RVSP - right ventricular systolic pressure, RVDP - right ventricular diastolic pressure, $+(dP/dt)_{max}$ - maximum rate of pressure development, CO - cardiac output. Values are means ± S.E.M. of six to eight experiments. Data from (4) with permission.

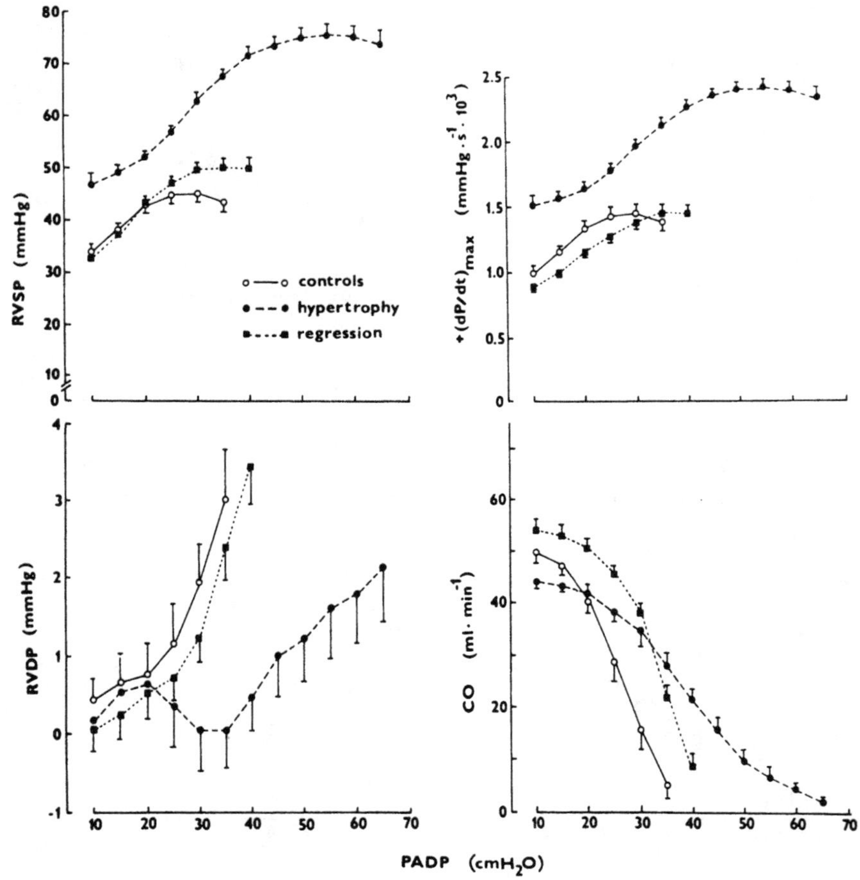

Fig. 3. Changes in the right ventricular functional parameters with an increase in pulmonary artery diastolic pressure (PADP, afterload) in hearts of IHA-exposed rats (hypertrophy) and of rats five weeks after the last hypoxic exposure (regression). Right atrial filling pressure was set at 10 cm H_2O. See legend to Fig. 2 for further description. Data from (4) with permission.

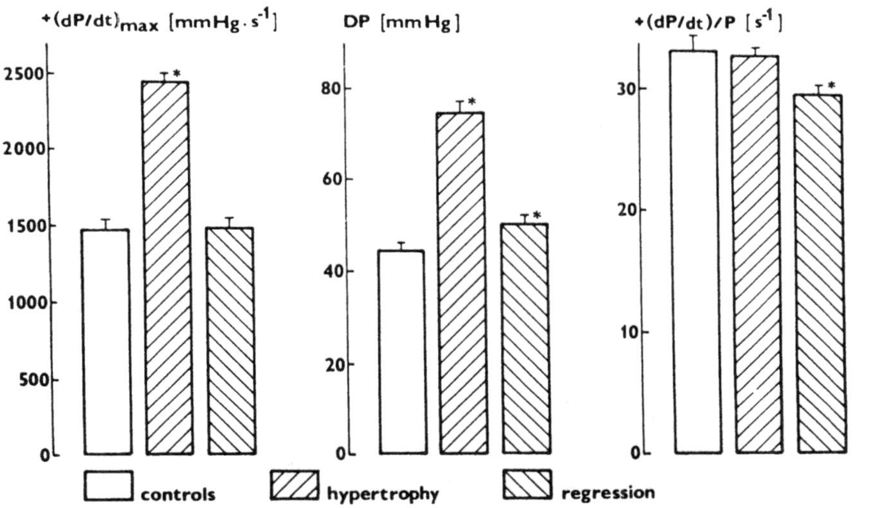

Fig. 4. Maximum functional performance of the right ventricle of IHA-exposed rats (hypertrophy) and of rats five weeks after the last hypoxic exposure (regression). $+(dP/dt)_{max}$ - maximum rate of pressure development in the right ventricle, reached at increased afterload, DP - peak right ventricular developed pressure, reached at increased afterload, $+(dP/dt)/P$ - ratio of $+(dP/dt)_{max}$ to DP (index of contractility). Values are means ± S.E.M. of six to eight experiments. * $P < 0.05$ vs. controls. Data from (4) with permission.

Effect of Tricuspid Incompetence

One of the causes of a gradual decline of cardiac output with increasing afterload could be a backward flow of medium during ventricular systole due to incompetent tricuspid valve resulting from the expansion of the tricuspid orifice. In fact, we indeed observed tricuspid regurgitation in our experiments at higher afterload as was evident from the backward flow into the filling circuit. In order to estimate a possible role of tricuspid regurgitation in assessment of the RV contractile performance a small glass-made valve was developed which is schematically drawn in Fig. 5. The valve consisted of an external tube jacket with an inner diameter of 2.4 mm and a length of 6 mm. The tube contained a moveable x-shaped rod 5.2 mm in length,

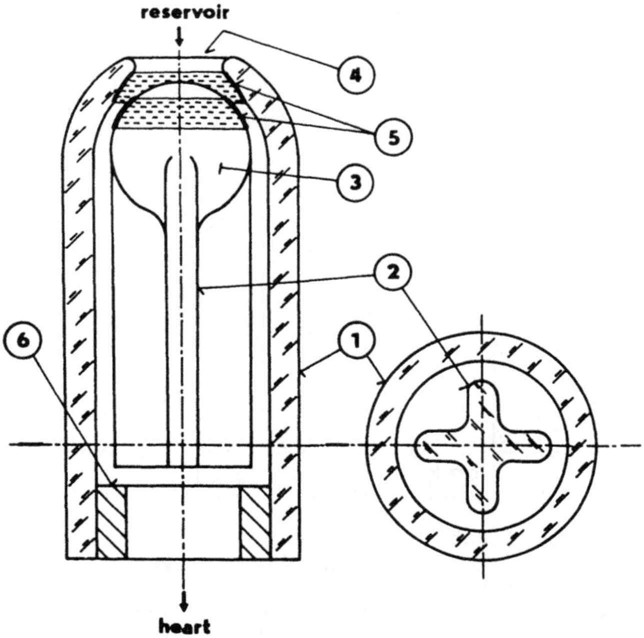

Fig. 5. The schematic representation of the artificial valve and its cross-section. 1 - external tube jacket, 2 - x-shaped rod, 3 - ball end of the rod, 4 - cone opening of the jacket, 5 - fine grinding of the rod and jacket opening, 6 - polyethylene tube stopper.

finely ground to a ball end of 1.8 mm diameter closely fitting into the 1.5 mm diameter cone opening of the jacket. The valve was inserted via the right atrium to the RV, tightened by a thread to the right atrium and connected to a filling reservoir placed 20 cm above the heart. While the RV filling was decreased but not prevented in this arrangement, the regurgitation did not occur. This was evident from the absence of a backward flow of medium to the filling reservoir. In another group of control hearts (prepared as described above), the artificial valve was omitted and the RAFP was set at zero level, so that the tricuspid incompetence resulting in the backward flow could occur. These ventricles could eject the coronary flow only. The RV function as a response to the gradual increase of PADP was compared in those two groups.

Typical recordings of the RV pressure response to the increased

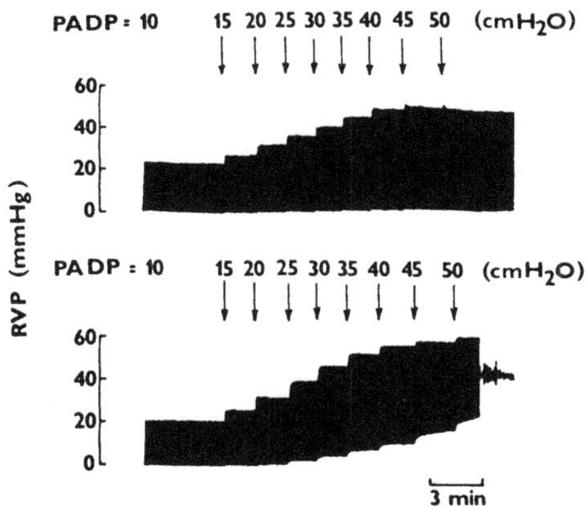

Fig. 6. Typical traces of the right ventricular pressure (RVP) changes with an increase of the pulmonary artery diastolic pressure (PADP) in the heart with the right atrial filling pressure set at zero level (upper trace) and in the heart with an artificial valve in the right ventricle (lower trace). Reprinted from (4) with permission.

afterload in hearts with an intact and artificial valve are illustrated in Fig. 6 and the data are summarized in Fig. 7. It is obvious that the tricuspid incompetence did not affect the maximum systolic performance; the peak $+(dP/dt)_{max}$ and RVDevP did not differ in the two groups. It indicates that the observed functional improvement of the hypertrophic RV in our model of chronic hypoxia appears to reflect the altered intrinsic ventricular performance rather than the altered function of the tricuspid valve. The hearts with the implanted artificial valve were able to maintain higher cardiac output at increased pulmonary outflow resistance. At the same time, the diastolic dysfunction (as evident from the dramatic rise of RVDP) became more pronounced than in the hearts with the intact valve. This suggests that the tricuspid regurgitation appears to be a very efficient mechanism protecting the RV against acute pressure overload.

Right Ventricular Function in Situ

The maximum contractile function of the RV was assessed by using an acute pressure load test in the open-chest in situ heart (Cihák et

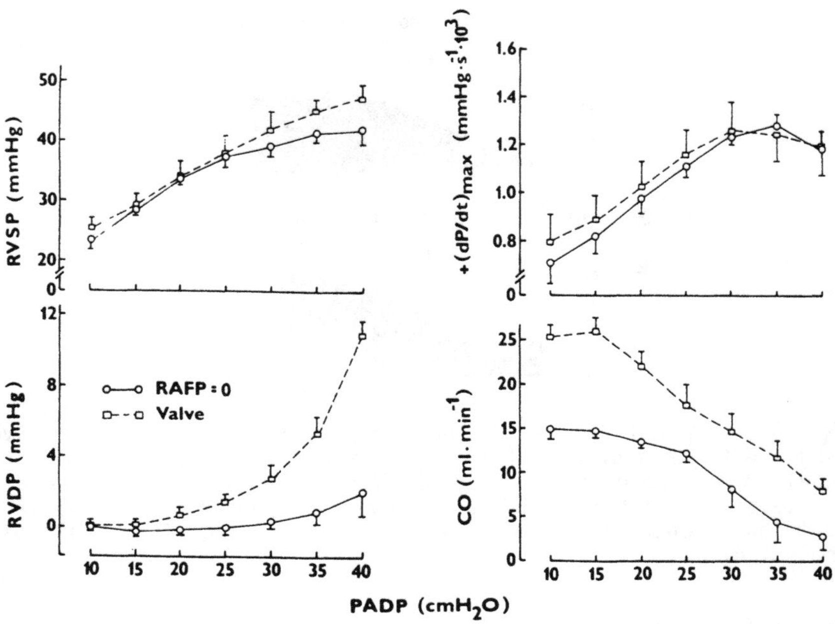

Fig. 7. Changes in the right ventricular functional parameters with an increase in pulmonary artery diastolic pressure (PADP) in hearts with the right atrial filling pressure set at zero level and in hearts with an artificial valve in the right ventricle. Values are means ± S.E.M. of four experiments in each group. See legend to Fig. 2 for further description. Data from (4) with permission.

al 1991). In anesthetized rat the trachea was cannulated and connected to a positive pressure ventilation pump (Zimmermann, 30 cycles.min^{-1}) and a middle sternotomy was performed. A ligature was prepared around the pulmonary artery and a 21-gauge needle was inserted into the RV cavity transmurally through the apex. The needle was attached via a cannula to the pressure transducer. The measuring and recording systems with an automated analysis of the parameters derived from the intraventricular pressure curve were generally the same as described in the previous section.

After stabilization of pressure (within 2 min) the ligature around the pulmonary artery was tight and the maximum response of the

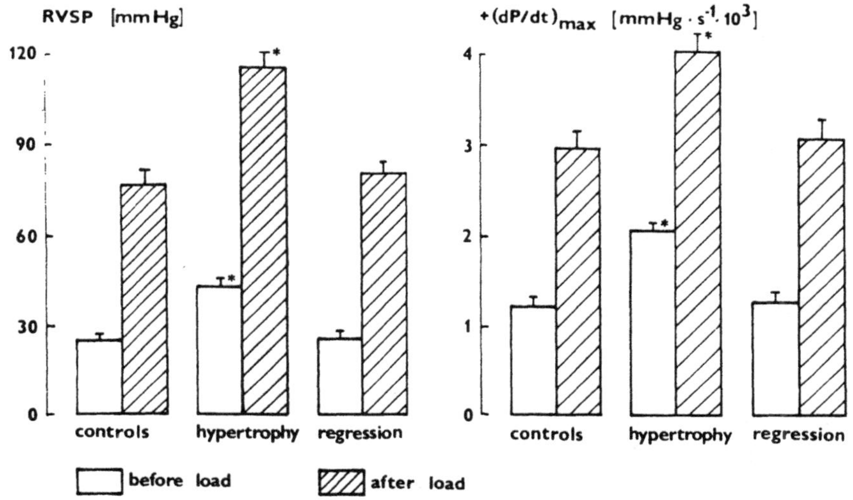

Fig. 8. The right ventricular systolic pressure (RVSP) and the maximum rate of pressure development [$+(dP/dt)_{max}$] in IHA-exposed rats (hypertrophy) and in rats five weeks after the last hypoxic exposure (regression) before and after acute ligation of the pulmonary artery. Values are means ± S.E.M. of eight to nine experiments. * $P < 0.05$ vs. controls.

RV to the acute pressure load was recorded. The functional reserve of the RV was defined as a difference between peak $+(dP/dt)_{max}$ after and before load.

Effect of IHA hypoxia

The resting values of RVSP measured in open chest rats by a direct piercing of the RV cavity were about 25 mm Hg and 43 mm Hg in control and hypertrophic groups, respectively, which is in a good agreement with the data obtained on intact animals (the same model) by introducing the catheter into the RV via the jugular vein (11). After loading (the acute ligation of the pulmonary artery) the RVSP increased by a maximum of 52 mm Hg in controls and by 73 mm Hg in hypertrophic ventricles. The resting value of the $+(dP/dt)_{max}$ of the hypertrophic RV was higher by 65% compared to the controls. After loading the peak increase of this parameter (functional reserve) was 1740 mm $Hg.s^{-1}$ in controls and 2030 mm $Hg.s^{-1}$ in the hypertrophic RV; the difference was not statistically significant (Fig. 8). As also is evident from Fig. 8, the regression of hypertrophy was associated with the complete reversal of both resting and peak values of the RVSP and $+(dP/dt)_{max}$.

Conclusions

We compared two methods modified for the assessment of the RV function using the rats with the RV hypertrophy induced by IHA hypoxia as the experimental model. The "in vitro" method was based on the isolated RV working heart preparation which makes it possible to assess ventricular contractile and pump performance under controlled loading conditions. The influence of the tricuspid valve incompetence was also tested in this preparation. The "in situ" method consists in the estimation of the maximum contractile function of the ventricle as a response to acute pressure load. Both methods appear to be suitable for the assessment of the RV function of small laboratory animals. While the first one offers the advantage to study both preload and afterload influences on the RV function with higher accuracy and discrimination, the second one represents a more simple screening test.

Both methods revealed markedly elevated performance of the hypertrophic RV in the IHA-exposed rats which serves to compensate the increased afterload without disturbing the pump activity. No signs of contractile dysfunction or pump failure were observed at this stage of hypertrophy. The regression of ventricular mass was accompanied by an almost complete reversal of ventricular function.

References

1. Ostádal B, Widimsky J. Intermittent Hypoxia and Cardiopulmonary System. Academia, Prague, 1985.
2. Anversa P, Beghi C, Levicky V, McDonald SL, Kikkawa Y. Am J Physiol 1982;243:H856-H861.
3. Zierhut W, Zimmer H-G. J Mol Cell Cardiol 1989;21:617-624.

4. Cihák R, Kolár F, Pelouch V, Procházka J, Ostádal B, Widimsky J. In: Right Ventricular Hypertrophy and Function in Chronic Lung Disease (eds.) V. Jezek, M Morpurgo, R Tramarin. Springer, Berlin, Heidelberg, New York 1991;in press.
5. Werchan PM, McDonough KH. Proc Soc Exp Biol Med 1987;185:339-345.
6. Kolár F, Ostádal B. Pfügers Arch 1991;419:121-126.
7. Neely JR, Liebermeister H, Battersby EJ, Morgan HE. Am J Physiol 1967;212:804-814.
8. Goodkind MJ. Circ Res 1968;22:605-614.
9. Saragoca MA, Tarazi RC. Hypertension 1981;3:(Suppl. II):171-176.
10. Pelouch V, Ostádal B, Procházka J, Urbanová D, Widimsky J. Prog Respir Res 1985;20:41-48.
11. Kol ár F, Ost ádal B, Proch ázka J, Pelouch V, Widimsky J. Respiration 1989;56:57-62.

Chronic Hypoxia-Induced Right Ventricular Enlargement: Age-Dependent Changes of Collagenous and Non-Collagenous Cardiac Protein Fractions

V. Pelouch, B. Ostádal, F. Kolář, M. Milerová[1], J. Grünermel

Institute of Physiology, Czechoslovak Academy of Sciences and
[1]Kardiocentrum, University Hospital Motol
Prague, CZECHOSLOVAKIA

Introduction

Exposure of experimental animals to chronic hypoxia induces pulmonary hypertension and right ventricular enlargement in a relatively short time period (for review see 1, 2). Cardiac enlargement may be due to an increased number of individual cell elements (hyperplasia) and their increased volume (hypertrophy) caused by increased protein synthesis. Participation of both processes in cardiac growth depends on cell type and age of the animal. Whereas connective tissue cells can proliferate during the whole ontogeny, the mitotic activity of ventricular myocytes is limited to a relatively short postnatal period; in rats it ceases 4-6 weeks after birth (3, 4).

It was the aim of the present study to establish whether the myocardial protein profile of rats exposed to chronic hypoxia just after birth, will be different from that of adult hypoxic animals. A particular attention was paid to possible hypoxia - induced qualitative and quantitative changes of the collagenous and non-collagenous proteins.

Material and Method

Animals and Experimental Procedure

Male Wistar rats were used. Intermittent high altitude hypoxia (IHA) was simulated in a low pressure chamber for 8h/day, 5 days a week. Barometric pressure (P_B = 306.8 mm Hg, PO_2 = 63.8 mm Hg, PCO_2 = 0.08 mm Hg) was reached after 13 exposures; the total number of exposures was 24. Two groups of rats were exposed to IHA: the first group ("young" - Y) already from the 4th day of postnatal life (8 sucklings per litter with mother), the second one ("adult" - A) after the 12th week of life. Two control groups ("young" and "adult") of animals were kept at normal P_B and PO_2 (equivalent to an altitude of 200 m) for the same time period. All animals had free access to water and were fed standard laboratory diet. After the last exposure, the animals were killed by decapitation and their hearts were dissected into the right ventricle, left ventricle and septum using the method of Fulton et al (5).

Protein Profiling

Right and left ventricles were washed with deionized water, cut into small pieces and briefly homogenized. The samples were treated consecutively with different buffers to obtain the fraction of sarcoplasmic and contractile proteins. The pellet was then treated with 0.5 M acetic acid and subsequently with 0.5 M acetic acid + pepsin to remove soluble collagenous proteins. The residue was treated with 1.25 M NaOH (45 min at 100 °C) to remove insoluble collagenous proteins. Proteins were determined according to Lowry et al (6). Concentration of the amino acid hydroxyproline in the collagenous fraction was estimated according to Huszár (7); all values were expressed as $mg.g^{-1}$ wet weight. Myosin was simultaneously isolated from the right and left ventricles of both the control and hypoxic animals; calcium ATPase activity of myosin was measured at 25 °C in 20 mM TRIS, 0.6 M KCl, 5.5 mM ATP in the presence of 10 mM $CaCl_2$ (pH = 7.5) (8), the isoenzyme myosin composition was established according to Hoh et al (9). Collagenous proteins were electrophoretically separated using the Hayashi procedure and the collagen I/collagen III ratio was calculated.

Ontogenetic Development of Myocardial Protein Composition

The transition from the neonatal period to adulthood is characterized by increasing circulatory demands. Whereas the volume load rises equally in both right and left ventricles, the increase in pressure load is significantly higher in the left heart. A faster growth of the left ventricle, with decreasing right to left ventricular weight ratio is a developmental consequence of this adaptive process.

Cardiac growth is accompanied by quantitative and qualitative changes of individual myocardial protein fractions. We have observed that in newborn rats the concentration of sarcoplasmic, contractile and collagenous proteins is the same in the right and left ventricles. During a further development, the amount of contractile and collagenous proteins in both ventricles increases. However, the increase of the contractile protein fraction is significantly higher in the left ventricle. Hence, significant right to left differences can be observed already on day 15 after birth; concentration of the contractile proteins is significantly higher in the left ventricle, whereas that of the collagenous fraction in the right ventricle and this difference persists even in adulthood. Substantially less pronounced developmental changes were detected in the fraction of sarcoplasmic proteins; their concentration increases only transiently in the right ventricle and this difference stepwise disappears.

Quantitative ontogenetic changes are accompanied by a qualitative remodelling of both contractile and collagenous proteins. Contractile proteins consist of two major components, myosin and actin and also include regulatory and modulatory proteins. Detailed developmental changes have so far been described predominantly in myosin (9, 10, 11, 12). The changes are associated with both the light and heavy chains and with phosphorylation of the LC 2 light chain. The rat myocardium is formed by three isoenzymes, V_1-V_3; the isoenzyme V_3 has the lowest ATPase activity (9). A newborn rat myocardium contains only V_3; during further development the synthesis of V_1 occurs and attains already 100% of the myosin

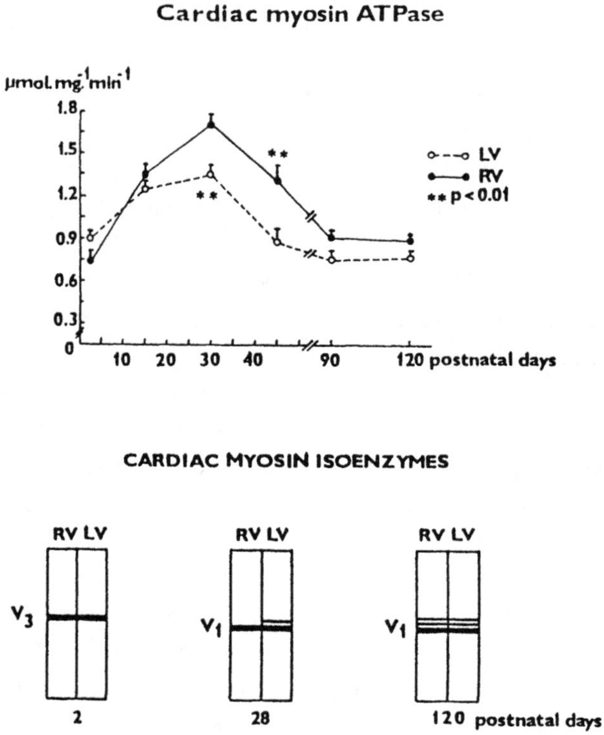

Fig. 1. Ontogenetic development (rat) of cardiac myosin ATPase and myosin isoenzymes in the left (LV) and right (RV) ventricles. Asterisks indicate significant differences between both ventricles; Student's t-test.

molecule at the time of weaning; consequently the ATPase activity rises substantially (13). Between the 30th and 60th day of life all three isoenzymes are synthesized and the ATPase activity declines (Fig. 1). The apparent right to left difference in the enzyme activity is probably due to the different relative proportion of isomyosins V_1 and V_3; the concentration of V_3 is significantly lower in the right ventricle (14). The actin activated ATPase of myosin and K-ATPase activity of this protein also change during development, during the first postnatal week in particular (12).

Collagenous proteins are formed by a mixture of different

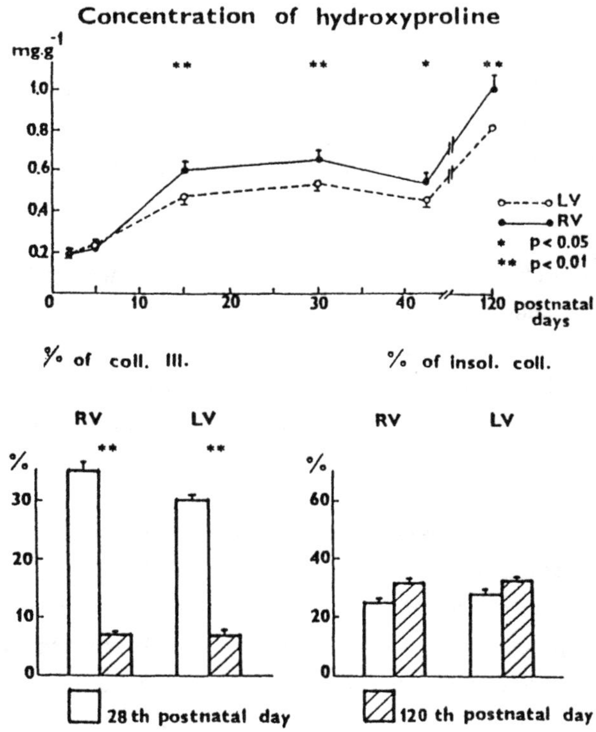

Fig. 2. Ontogenetic development (rat); concentration of hydroxyproline (upper part), percentage of collagen III and percentage of insoluble collagenous proteins (lower part) in the left (LV) and right (RV) ventricle. Asterisks indicate significant differences; Student's t-test.

collagens (type I-V), elastin, glycoproteins and glycosaminoglycans (15). The collagenous matrix is formed primarily by collagen I that aggregates into thick fibers and by collagen III (thin fibers); collagen IV is present in both fibroblasts and myocytes; the functional significance of the heterogeneity of cardiac collagen types is still a matter of speculations. The neonatal period is characterized by a relatively high concentration of collagen III; in adulthood the relative proportion of this fraction decreases to about 20% (15) (Fig. 2).

It may be concluded that the myocardium is composed of material formed by muscular and interstitial compartments; their relative proportion which determines the myocardial function, changes significantly during ontogeny. The precise knowledge of this developmental process is prerequisite for providing a framework within which the effect of different pathophysiological situations - e.g. chronic hypoxia - can be elucidated.

Age-Dependent Effect of Chronic Hypoxia on the Cardiac Proteins

IHA induced a chronic pulmonary hypertension and right ventricular hypertrophy in rats of both age groups (16). Whereas the pressure elevation was more expressed in adult hypoxic animals, the right ventricular enlargement was higher in the young group. This result suggests that the participation of the increased pressure load on the ventricular growth response may differ during development. The close relationship between the degree of pulmonary blood pressure and right ventricular weight was found only during the early phases of ontogenetic development. In this period the mitotic activity of ventricular myocytes can still be observed. Therefore, when exposed to hypoxia they increase not only in size but also in number (17, 4). The stimulus may be the decrease of PO_2 and/or the secondary effect of altered haemodynamics (18). The steep relationship between pulmonary blood pressure and right ventricular weight, as well as the close correlation between these two variables in young hypoxic animals thus might be the consequence of a higher "reactivity" of the developing myocardium, in which both the above mentioned mechanisms leading to the ventricular enlargement may be involved (16).

IHA induced significant changes in protein profiling of the myocardium in both age groups. The myosin ATPase activity in the hypertrophic right ventricle decreases together with the increase of proportion of isomyosin V_3 (Fig. 3). Molecular mechanism responsible for changes in enzymatic activity of cardiac myosin is age-dependent, whereas in young animals, IHA stimulates the shift from V_1 to V_3 isomyosin; in adult hypoxic animals it increases the synthesis of isomyosin with a lower ATPase activity (19).

IHA also modulates the qualitative and quantitative changes of collagenous proteins (Fig. 4). In both age groups, a significant

Table 1.

	Controls		IHA	
	RV	LV	RV	LV
Young	1.18 ±0.02	0.47 ±0.01	0.25 ±0.02	0.83 ±0.05
Adult	1.31 ±0.03	1.25 ±0.02	1.15 ±0.04	1.12 ±0.01

Collagen I/collagen III ratio in controls and IHA - exposed animals. RV - right ventricle, LV - left ventricle; means ± S.E.M.

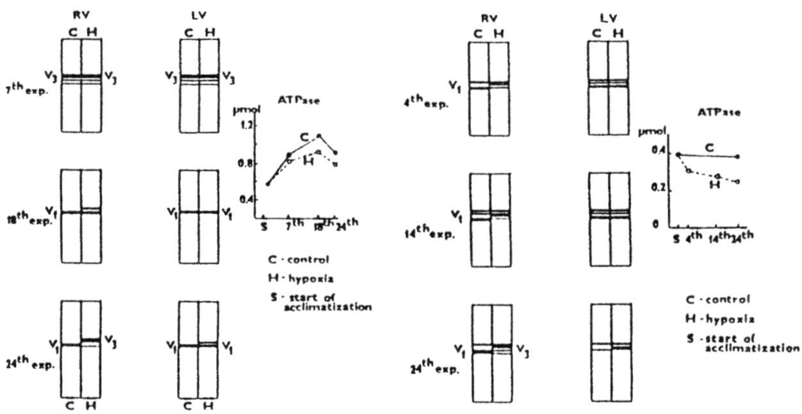

Fig. 3. Myosin isoenzymes and myosin ATPase activity in the right (RV) and left (LV) ventricles of the controls (C) and IHA - exposed (H) rats. Left: acclimatization from the 4th postnatal day, right: acclimatization from the 12th week.

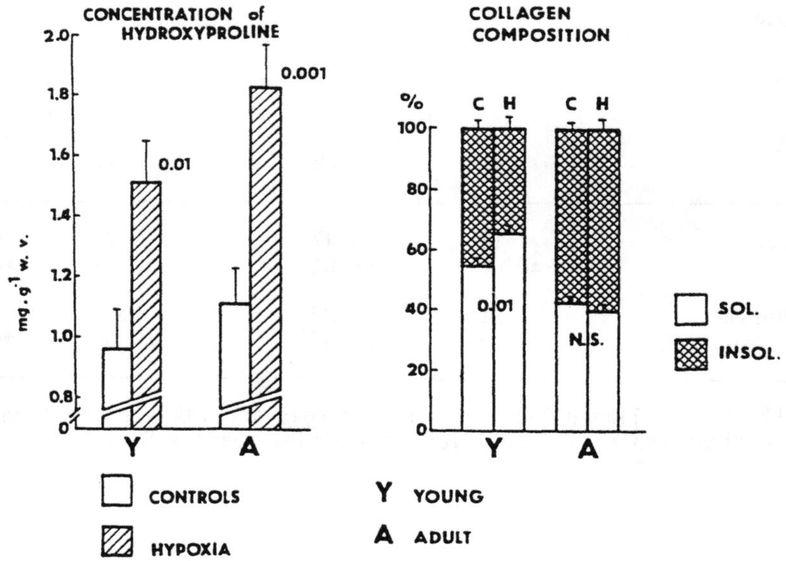

Fig. 4. Collagen in the right ventricle. Concentration of hydroxyproline and percentage of soluble (sol) and insoluble (insol) collagenous proteins in controls and IHA - exposed rats.
Y - acclimatization from the 4th postnatal day (young)
A - acclimatization from the 12th postnatal week (adult)

increase in hydroxyproline concentration can be observed; this effect was, however, more pronounced in animals exposed to IHA in adulthood. In young animals the elevation of hydroxyproline is predominantly due to the increase of concentration of soluble collagenous proteins; in adult animals the proportion of both fractions remains unchanged. IHA decreases the collagen I/collagen III ratio more in young animals; this change suggests that the synthesis of collagen III significantly increases (Table 1).

It may be assumed that IHA significantly affects protein profiling of the cardiac muscle. The relative proportion of contractile and collagenous proteins, together with their qualitative changes may significantly influence not only contractility of the hypoxic myocardium but probably also its resistance to acute lack of oxygen.

Conclusions

Chronic hypoxia induced significant quantitative and qualitative changes of cardiac proteins in young as well as in adult animals. The shift of cardiac isomyosins from V_1 to V_3, together with a decrease of myosin ATPase activity was observed in both age groups. The synthesis of the collagenous protein fraction was more expressed in adult hypoxic animals; moreover, the collagenous fraction in these rats contains a higher amount of insoluble collagen. On the other hand, the collagen I/collagen III ratio decreased more in young hypoxic animals suggesting an increased synthesis of collagen III in this age group. All these changes have to be taken into consideration for the explanation of age-dependent contractile properties of the hypoxic heart.

References

1. Fishman AP. Circulation Res 1976;38:221-231.
2. Ostádal B, Widimsky J. Intermittent hypoxia and cardiopulmonary system. Academia, Prague, 1985.
3. Zak R. Circulation Res 1974;35(Suppl II):17-27.
4. Wachtlová M, Mares V, Ostádal B. Wirchows Arch B Cell Pathol 1977;24:335-342.
5. Fulton RM, Hutchinson RM, Jones AM. Brit Heart J 1952;14:413-420.
6. Lowry OH, Rosenbrough NJ, Farr AL. J Biol Chem 1951;193:265-273.
7. Huszar G. Anal Biochem 1980;105:424-429.
8. Pelouch V, Ostádal B, Urbanová D, Procházka J, Ressl J, Widimsky J. Physiol Bohemoslov 1980;29:313-322.
9. Hoh JFY, McGrath PA, White RI. J Mol Cell Cardiol 1978;10:1053-1076.
10. Medugorac I, Hoppe-Seylers Z. J Physiol 1979;360:326-330.
11. Lompre AM, Schwartz K, d'Alibis A, Lacombe G, vanThien N, Swynghedauw B. Nature 1979;282:5-12.
12. Watras J. J Mol Cell Cardiol 1981;13:1011-1021.
13. Pelouch V, Milerová M, Ostádal B, Procházka J. In: Right ventricular hypertrophy and lesser circulation (eds.) Marpurgo

A, Jezek V. 1991; (in press).

14. Brooks WW, Bing OHL, Blaustein AS, Allen PD. J Mol Cell Cardiol 1987;19:433-440.
15. Weber K. J Am Coll Cardiol 1989;13:1637-1652.
16. Kolár F, Ostádal B, Procházka J, Pelouch V, Widimsky J. Respiration 1989;56:57-62.
17. Hollenberg M, Honbo N, Samorodin AJ. Am J Physiol 1976;231:1445-1450.
18. Dowell RT, Martin AF. Am J Physiol 1984;247:H967-H972.
19. Pelouch V, Ostádal B, Procházka J, Urbanová D, Widimsky J. Prog Resp Res 1985;20:41-48.

Sarcolemmal Cation Transport Systems in Rat Hearts Acclimatized to High Altitude Hypoxia: Influence of 7-Oxo-Prostacyclin

A. Ziegelhöffer[1], J. Grünermel[2], A. Dzurba[1], J. Procházka[2], F. Kolář[2], N. Vrbjar[1], V. Pelouch[2], B. Ostádal[2], L. Szekeres[3]

[1]Institute for Heart Research, Slovak Academy of Sciences, Bratislava, CZECHOSLOVAKIA; [2]Institute of Physiology, Czechoslovak Academy of Sciences, Prague, CZECHOSLOVAKIA; [3]Institute of Pharmacology, Albert Szent-Györgyi University, Szeged, HUNGARY

Introduction

Acclimatization to intermittent high altitude (IHA) is characterized by the development of polycythemia, pulmonary hypertension and right ventricular hypertrophy. Furthermore, at the beginning of the adaptation process, necrotic lesions can be observed, localized predominantly in the right ventricle (1, 2). In this connection, it is interesting to note that such an affected myocardium exhibits a significantly increased resistance to acute anoxia "in vitro", as well as to isoproterenol-induced myocardial lesions (3, 4). Recently, we have observed that in hearts acclimatized to IHA the sarcolemmal Na, K-ATPase, Mg-ATPase and low affinity Ca-ATPase exhibited decreased V_{max} and K_m values; the latter being considered as a sign of adaptation to hypoxia at the enzyme level (5).

7-oxo-prostacyclin (PGI_2), was reported to protect the heart against ischemia (6) postischemic reperfusion damage (7) and arrhythmias of ischemic and non-ischemic origin (8). These late appearance and long lasting actions of PGI_2 were attributed to its membrane-stabilizing effect. The molecular mechanism of PGI_2 action involves an increase in the sarcolemmal Na, K-ATPase activity (9, 10), via both the enzyme induction and increased affinity of its active site towards ATP (11). The aims of the present study were:

1. to evaluate the antianoxic cardioprotective effect of PGI_2 in control and IHA-acclimatized animals;

2. to test whether PGI_2 can affect, besides the activity of Na, K-ATPase, also the activities and properties of other cardiac sarcolemmal ATPases, such as Mg-ATPase and Ca-ATPase with low affinity for calcium with a special attention to the possible right and left ventricular differences;

3. to check whether PGI_2 may also be similarly effective in hearts adapted to IHA, i.e. in hearts with right ventricular hypertrophy.

Material and Methods

A total of 80 male Wistar rats (70-day-old) were exposed to a simulated altitude in a barochamber, 4 h per day, 5 days a week. The acclimatization was stepwise so that the altitude of 7000 m (barometric pressure 40.9 kPa, PO_2 8.5 kPa, PCO_2 0.01 kPa) was reached after 13 exposures; the total number of exposures was 32. Experimental animals had free access to water and were fed standard laboratory pellet diet. Control animals were kept under normoxic conditions for an identical time period.

Immediately after the last hypoxic exposure the animals received a single dose of 5 ug PGI_2 in 100 ul per 100 g body weight, intramuscularly; control animals received the same volume of saline. Twenty four hours after the last exposure to IHA (and administration of PGI_2) the animals were killed by decapitation. Hearts were excised quickly by means of the method of Fulton et al (12) recommended by the WHO expert committee (13) and dissected into three parts, the right and left ventricles and septum which were weighed separately. The ventricles were then preserved in liquid nitrogen until the isolation of the sarcolemmal fraction.

In some hearts of each experimental group the papillary muscles were prepared and mounted in a perfusion device with modified Tyrode buffer containing (in $mmol.l^{-1}$): NaCl 145; KCl 5.9; $MgCl_2$ 1.2, $CaCl_2$ 2.5; glucose 11 and HEPES 5.0 (pH 7.4). Muscles were stimulated by constant suprathreshold rectangular pulses at a rate of 150 per min. The medium was gassed with pure oxygen; for the anoxic period of 120 min oxygen was replaced by nitrogen. Subsequent to anoxia, the muscles were reoxygenated with a normoxic medium for additional 60 min. The tolerance of the myocardium to acute anoxia was evaluated according to the maximum contractile force obtained during reoxygenation and expressed as percentage of a steady-state preanoxic value. Isolated membrane preparations enriched in sarcolemma were obtained from separated left and right ventricles using the method of hypotonic shock (14) modified as described previously (5). The resulting fractions consisted predominantly of right-side-out membrane vesicles. Contamination with other subcellular membrane particles was below 3% as measured by specific activities of 5'-nucleotidase, succinic dehydrogenase, oligomycin-sensitive Mg-ATPase and Mg-dependent, Ca-stimulated ATPase (15, 16).

Specific activities of the sarcolemmal Na, K-ATPase, Mg-ATPase and Ca-ATPase with low affinity for calcium (17) were estimated by measuring the amount of phosphate liberated from ATP splitting. The reaction was allowed to proceed for 10 min at 37 °C in 1 ml of a medium containing 50 $mmol.l^{-1}$ histidine-HCl buffer pH = 7.0, 30-56 ug of membrane protein and an optimal composition of cationic ligands as follows: 2, 10 and 100 $mmol.l^{-1}$ $MgCl_2$, KCl and NaCl for Na, K-ATPase; 2 $mmol.l^{-1}$ $MgCl_2$ or $CaCl_2$ for Mg-ATPase or Ca-ATPase, respectively. The enzyme reaction was started by the addition of (2 $mmol.l^{-1}$, final concentration) ATP and terminated by 1 ml of ice-cold trichloroacetic acid (0.73 $mol.l^{-1}$). The activity of Na, K-ATPase was given by the difference in the enzyme activity in the presence of optimal stimulatory concentrations of Na and K ions. Approximately 70% of the enzyme activity was sensitive to ouabain (1 $mmol.l^{-1}$). Further details concerning estimation of the enzyme activities have been described elsewhere (15).

The techniques applied in studies on the effect of cycloheximide on the induction of sarcolemmal Na, K-ATPase by PGI_2, as well as in estimation of parameters of enzyme kinetics by means of computer-assisted non linear curve fitting, have already been described (11).

Results and Discussion

Some selected weight characteristics of control animals and those acclimatized to IHA are presented in Fig. 1. IHA induced a significant decrease of body weight as compared with control animals. The relative right ventricular weight, as well as the right to left ventricular weight ratio were significantly increased, indicating a marked hypertrophy of the right ventricle. The enhanced LVW/BW ratio also indicates a certain degree of the left ventricular hypertrophy.

Our earlier results showed that acclimatization to IHA increases the capability of the heart to tolerate acute anoxia "in vitro" as judged from the faster and more complete recovery of contractility after a period of transient anoxia (5). The results given in Fig. 2 show clearly that the administration of PGI_2 (6, 7, 17) increases significantly the tolerance of the myocardium to acute anoxia, which is similar to the acclimatization to IHA (1-5). Both effects seem to be independent and additive. This supports the view that, at least in the heart, acclimatization to IHA hypoxia may be considered as a special kind of preconditioning; the effect on recovery of contractile force induced by administration of PGI_2 to animals preconditioned by acclimatization to IHA represents the sum of the effects observed with PGI_2 and IHA separately.

The molecular mechanism of the late effect of PGI_2 on the heart was already partly revealed. It was demonstrated that the drug acts via stabilization of the cell membrane function (9, 18, 19). The latter process involves enhancement of the Na, K-ATPase activity (Table 1) accompanied by a decreased K_m value, i.e. an increased affinity of the enzyme active site to ATP (11).

A special reason for investigation of properties and activity of the sarcolemmal ATPases (and particularly that of the Na, K-ATPase) involved in the maintenance of cation homeostasis in cardiac cells is their capability of "adaptation" to lack of the substrate, i.e. ATP. It was found that after prolonged or repeated brief periods of hypoxia or ischemia, the activity of Na, K-ATPase (20), decreases

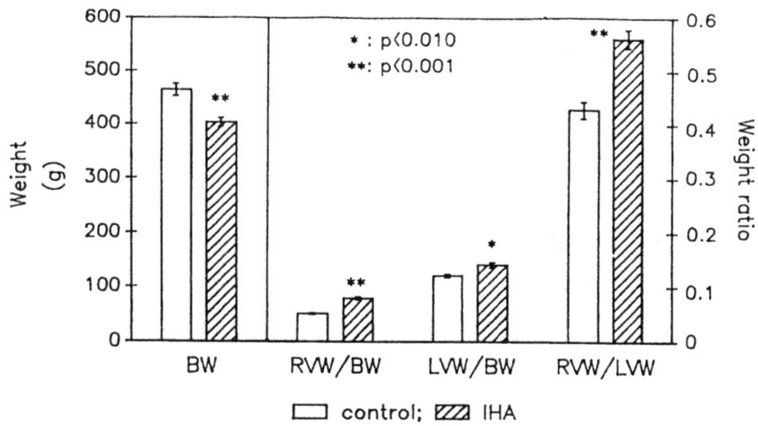

Fig. 1. Weight characteristics of rats acclimatized to IHA. Results are means ± S.E.M.; n = 12; statistical significance was established by means of the Student's t-test; $p < 0.01$. BW - body weight, RVW - right ventricular weight, LVW - left ventricular weight, RVW/LVW - left to right ventricular weight ratio.

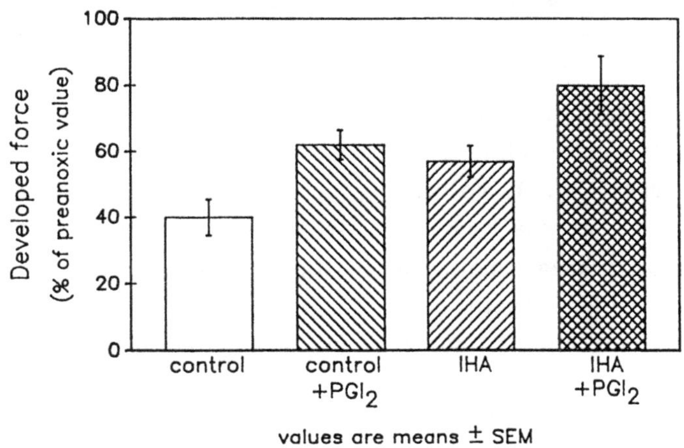

Fig. 2. Effect of PGI_2 on postanoxic force recovery in papillary muscle from hearts acclimatized to IHA and controls. Results are means ± S.E.M.; n = 12; statistical significance with respect to controls (Student's t test) - $p < 0.01$ in all groups. Abbreviations as in the text.

similar to the activities of Mg-ATPase and Ca-ATPase with low affinity to calcium decrease (5). Whatever the real cause of the above decrease of the actual ATPase activities might be, the investigation of their kinetic features revealed both decreased V_{max} and K_m values. From the point of view of membrane biochemistry and physiology, the above findings indicate that these enzymes are partially able to maintain their vital functions by lowering their substrate - demands (a lower ATP concentration is required for a half-maximal stimulation of the enzyme) via more effective utilization of the ATP available. In fact, it is well known that the membrane bound enzymes operate efficiently at substrate concentrations near their K_m values; hence, it appears that in ischemia or hypoxia the sarcolemmal ATPases may run on a spare regimen. Although it has not yet been discussed in the literature, this principle of "enzyme adaptation" may also be utilized most probably also in the preconditioning.

The results in Figs. 3, 4 and 5 demonstrate higher activities of the Na, K-ATPase. Mg-ATPase and low affinity Ca-ATPase in the left ventricle in comparison to the right ventricle of the rat heart. This nonhomogeneity of the enzyme apparatus of the sarcolemma may be related to the metabolic and structural heterogeneity of functionally different parts of the heart (21).

Our findings also confirmed that acclimatization to IHA decreases the activities of sarcolemmal Na, K-ATPase (Fig. 3, column 3), Mg-ATPase (Fig. 4, column 3), as well as the Ca-ATPase (Fig. 5, column 3) and, at the same time, it increases their affinity to ATP (see Ref. 5). When running on economy regimen adapted to decreased energy production, these enzymes may secure a degree of membrane function sufficient to support and even to improve recovery of contractility after anoxia (see Fig. 2).

It was already discussed that the molecular mechanism of the late action of PGI_2 involves, in addition to enzyme induction-related increase of the Na, K-ATPase activity (5, 22), an increased affinity of the enzyme to ATP (5). Since this modulation of affinity to ATP is similar to that induced by acclimatization to IHA, it was reasonable to investigate the effect of PGI_2 on acclimatized

Table 1. Heart sarcolemmal Na, K-ATPase activities and K_m values after i.m. application of a single dose of PGI_2 in a presence or absence of cycloheximide.

	umol. $P_i mg^{-1}.h^{-1}$	% Activity	K_m
Control	8.01 ± 0.29	100.0	81 ± 0.16
Control with cycloheximide	8.75 ± 0.03	100.2	0.77 ± 0.14
PGI_2	^{a}13.18 ± 0.73	164.5	^{a}0.50 ± 0.80
PGI_2 with cycloheximide	^{bc}9.56 ± 0.34	119.4	0.81 ± 0.08

Means ± S.E.M.; n = 5; a - $p < 0.001$ and b - $p < 0.01$ against the controls, significance in the presence compared to the absence of cycloheximide c - $p < 0.005$. Data taken from (11).

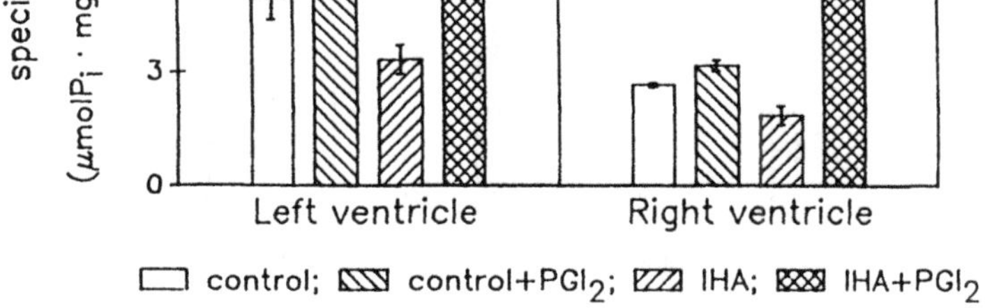

Fig. 3. Effect of PGI_2 on the activity of sarcolemmal Na, K-ATPase in hearts of rats acclimatized to IHA and controls. Results are means ± S.E.M., n = 6-10. Statistical significances (Student's t-test) $p < 0.001$ for control + PGI_2 with respect to the control in the LV and for IHA + PGI_2 in the RV; $p < 0.05$ for all other groups as related with the controls in the LV and RV.

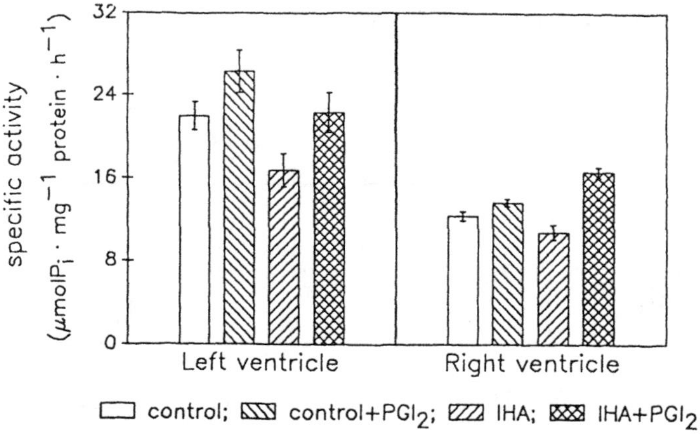

Fig. 4. Effect of PGI_2 on the activity of sarcolemmal Mg-ATPase in hearts of rats acclimatized to IHA and controls. Results are means ± S.E.M., n = 6-10; Statistical significances (Student's t-test): $p < 0.001$ for IHA in the LV and for IHA + PGI_2 in the RV with respect to controls; $p < 0.05$ for control + PGI_2 in the LV as well as for control + PGI_2 and IHA in the RV as related with the respective controls.

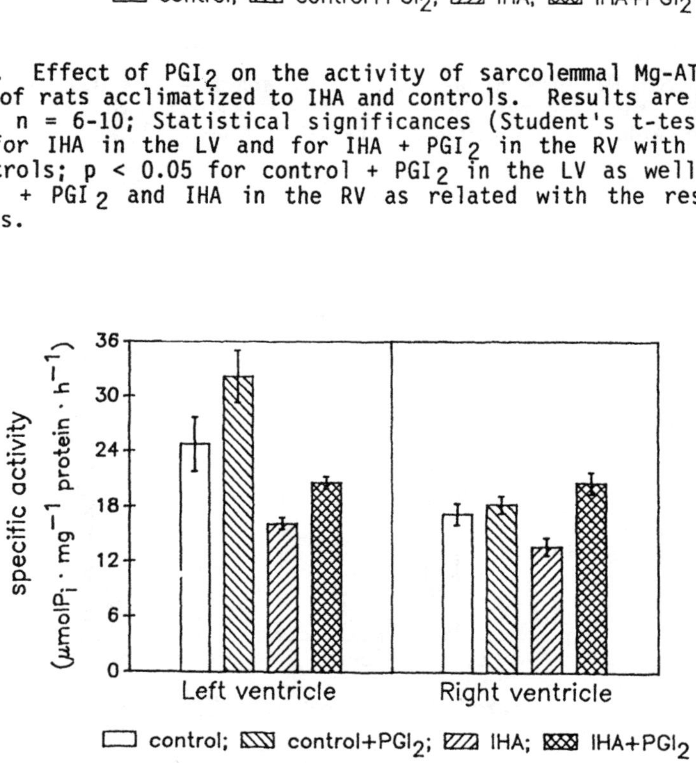

Fig. 5. Effect of PGI_2 on the activity of sarcolemmal Ca-ATPase with low affinity to calcium in hearts of rats acclimatized to IHA and controls. Results are means ± S.E.M., n = 6-10; Statistical significances (Student's t-test): $p < 0.001$ for IHA with respect to control in the LV as well as for IHA and IHA + PGI_2 with respect to control in the RV.

hearts. The results obtained (Figs. 3, 4 and 5) indicate that: i) the induction of Na, K-ATPase by PGI_2 may take place in spite of the fact that protein synthesizing mechanisms have already accelerated. This is manifested by an increased ATPase and particularly the Na, K-ATPase activities in both the left and right ventricles from hearts of acclimatized animals; ii) in contrast to the effect of acclimatization to IHA, the administration of PGI_2 increased the activity of sarcolemmal Na, K-ATPase. Nevertheless, this elevation did not depress but, on the contrary enhanced the tolerance of the heart against anoxia (see Fig. 2) as demonstrated with isolated superfused papillary muscles from acclimatized and PGI_2-treated animals.

Conclusions

The activities of sarcolemmal Na, K-ATPase, Mg-ATPase and low affinity Ca-ATPase are different in the left and right ventricles of the rat heart. The activities of these enzymes decrease and increase as a consequence of acclimatization to IHA and of late action of PGI_2, respectively. The effect of IHA on the properties of membrane bound ATPases may be indicated as adaptation on the enzyme level. It seems that PGI_2 and acclimatization to IHA represent two independent and additive cardioprotective effects against hypoxia.

References

1. Widimsky J, Urbanová D, Ressl J, Ostádal B, Pelouch V, Procházka J. Cardiovasc Res 1973;7:789-808.
2. Widimsky J, Ostádal B, Urbanová D, Ressl J, Procházka J, Pelouch V. Chest 1980;77:383-389.
3. Poupa O, Krofta K, Proch ázka J, Turek Z. Fed Proc 1966;25:1243-1246.
4. McGrath J, Procházka J, Pelouch V, Ostádal B. J Appl Physiol 1973;34:289-293.
5. Ziegelhöffer A, Procházka J, Pelouch V, Ostádal B, Dzurba A, Vrbjar N. Physiol Bohemoslov 1987;36:403-415.
6. Takáts I, Szekeres L. Acta Physiol Hung 1985;66:390.

7. Udvary E, Végh A, Szekeres L. Pharmacol Res Commun 1988;20:171-172.
8. Szekeres L, Szilvássy Z, Udvary E, Végh A. Pharmacol Res Commun 1988;20(Suppl I):77-78.
9. Szekeres L, Németh M, Szilvássy Z, Tősaki A, Udvary E, Végh A. Biomed Biochim Acta 1988;47:6-7.
10. Szekeres L, Koltai M, Pataricza G, Takáts I, Udvary E. Biomed Biochim Acta 1987;43:135-142.
11. Dzurba A, Ziegelöffer, A Breier A, Vrbjar N. Cardioscience 1991;2:105-108.
12. Fulton RN, Hutchinison EC, Jones AN. Brit Heart J 1952;14:413-420.
13. Chronic cor pulmonale. WHO report of an expert committee. World Health Organization Technical Report Series, No. 213, Geneva, 1961; pp. 92.
14. Dhalla NS, Lee SL, Anand MB, Chauhan MS. Biochem Pharmacol 1977;26:2050-2060.
15. Fedelesová M, Dzurba A, Ziegelhöffer A. In: Recent Advances of Studies on Cardiac Structure and Metabolism, Vol. 9. The Sarcolemma (eds.) PE Roy, NS Dhalla. University Park Press, Baltimore, Maryland, 1976; pp. 303-309.
16. Ziegelhöffer A, Breier A, Dzurba A, Vrbjar N. Gen Physiol Biophys 1983;2:447-456.
17. Ziegelhöffer A, Dhalla NS. J Mol Cell Cardiol 1979;11(Suppl II):69.
18. Szekeres L, Udvary E, Végh A. J Mol Cell Cardiol 1985;17(Suppl III):Abstract No. 129.
19. Szilvássy Z, Udvary E, Szekeres L, Végh A. J Mol cell Cardiol 22(Suppl III):PT 17.
20. Vrbjar N, Slezák J, Ziegelhöffer A, Tribulová N. Eur Heart J, 1991; in press.
21. Slezák J, Tribulová N. J Mol Cell Cardiol 1990;22(Suppl III):PF32, s.66.
22. Stankovicová T, Zemková H, Ziegelhöffer A, Vyskocil F. Brit J Pharmacol 1991; in press.

Effects on the Heart of Two Forms of Chronic Hypoxia

J. McGrath

Department of Physiology, Texas Tech University,
Health Sciences Center, Lubbock, Texas 79430 U.S.A.

Introduction

The physiological effects of chronic exposure to high altitude or carbon monoxide (CO) have been studied intensively in humans and animals, but the effects on the heart of these stressors combined have not been investigated. This research was initiated because theoretical and experimental data suggest that chronic exposure to CO combined with altitude may be more detrimental to the heart than chronic exposure to CO alone and because there are no data on the effects of chronic exposure to CO at altitude.

Methods

Exposure System

The altitude chamber system used in these studies has been described previously (1). It consists of six cylindrical chambers and a system of interconnected valves which maintains the separate chambers at either ambient barometric pressure or low barometric pressure (simulated high altitude). Air, supplied from a central air duct, enters the system through a HEPA filter and flows through each chamber at 55 l/min (16 air changes/h). In the altitude studies, the pressure in the chambers is reduced by a water-sealed pump (Atlantic Fluidic, Inc.) and changes in pressure are measured with an altimeter. In the altitude CO studies, the CO is provided in cylinders and introduced into the chambers through a multiple mass flow controller (Matheson Gas, Inc.). A switching solenoid (Chronotrol-Linberg Enterprises, Inc.) allows the air in each chamber to be sampled and analyzed in sequence. CO concentrations are measured with an infrared analyzer (Beckman Instruments, Inc) and recorded on a chart recorder (Houston Instruments, Inc). Chamber CO_2 concentrations are measured with an LB-2 CO_2 analyzer (Sensormedics, Inc.) The average ambient temperature and relative humidity for these studies were 22 ± 1°C and 50 ± 5%, respectively.

Exposures

Male laboratory rats, 6 animals to a cage, were placed in the chambers and exposed for 6 weeks to ambient altitude (AMB), 10,000 ft, 15,000 ft or 18,000 ft simulated high altitude, and carbon monoxide (CO) concentrations of 35-500 ppm. Exposures were continuous, except the chambers were opened to replenish food and water, clean cages, weigh animals and measure hematocrit ratios and COHb levels. The animals were fed a commercial diet and provided tap water ad libitum. The animals gained weight progressively throughout the exposure. Only in the animals exposed to altitudes equal to or greater than 15,000 ft was there a significant increase in hemoglobin concentration and decrease in body weight. These data were reported earlier (1,2).

Analytical Methods

Each week the animals were removed from the chambers and carboxyhemoglobin (COHb) and hematocrit ratios concentrations were measured immediately on blood samples obtained by snipping the tails; then the animals were weighed. COHb concentrations were determined spectrophotometrically (3,4) or by gas chromatography (5). COHb concentrations measured at the end of the six-week exposure are reported in Table 1.

After 6 weeks, the animals were killed and the hearts were removed and the atria, great vessels and visible fat were trimmed away. The right ventricles were dissected away and the left ventricles were opened and rinsed. Both ventricles were blotted dry on filter paper (Whatman no. 4) and weighed on a precision balance;

Table 1. Percent carboxyhemoglobin levels corresponding to given chamber carbon monoxide concentrations and altitudes[a]

	Carbon Monoxide Concentration					
Alt(ft)	0 ppm	9 ppm	35 ppm	50 ppm	100 ppm	500 ppm
3,300	0.6±0.1	0.9±0.1	2.4±0.2	3.7±0.1	8.5±0.4	40.0±0.4
10,000	1.3±0.2	1.8±0.2	3.3±0.2	ND[b]	9.4±0.4	ND
15,000	1.7±0.8	2.1±0.1	ND	ND	10.0±0.5	41.5±0.5
18,000	1.9±0.5	ND	ND	5.0±0.4	10.2±0.7	42.0±0.7

[a]Values are means ± SD. [b]ND = not determined.

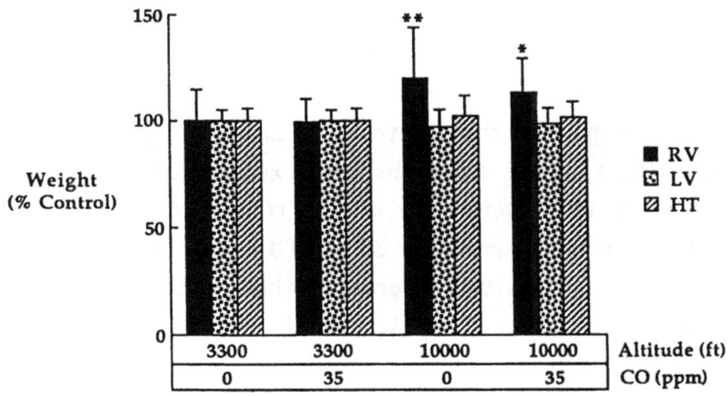

Fig. 1. Right ventricle (RV), left ventricle (LV) and Total Heart (HT) weights in rats exposed to 3,300 or 10,000 ft and/or 35 ppm CO. *$p < 0.05$, ** $p < 0.005$.

the weights of the right ventricles (RV), left ventricles plus septa (LV), and total heart (HT) were recorded.

Results

COHb levels in rats exposed to various combinations of CO

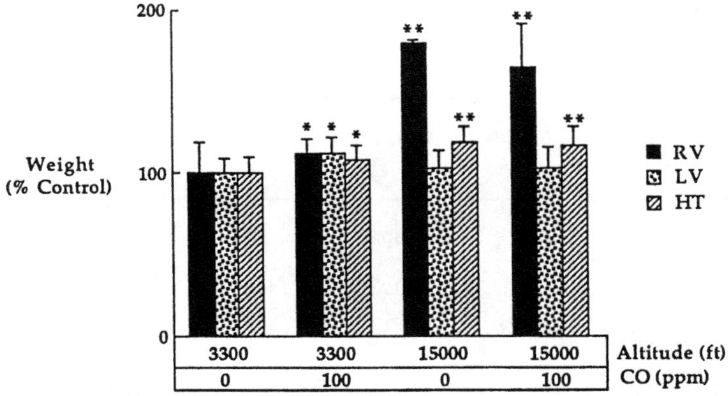

Fig. 2. Heart weights in rats exposed to 3,300 or 15,000 ft and/or 100 pmm CO. Legend as in Fig. 1.

increase with increasing exposure to CO (Table 1). There is a small increase in CO arising from the slightly higher basal COHb levels associated with increasing elevation which, in the absence of CO exposure (0 ppm), increases from 0.6% at 3,3000 ft to 1.9% at 18,000 ft altitude.

CO (35 ppm) had no effect on RV weights (Fig. 1). RV weights were significantly greater than controls in animals exposed to 10,000 ft (3,048 m). The altitude effect was not exacerbated by concomitant exposure to 35 ppm CO. There were no significant differences in LV or HT weights in animals exposed to 35 ppm CO and/or to 10,000 ft.

RV weights were significantly greater than controls (Fig. 2) in animals exposed to 100 ppm CO or 15,000 ft (4,572 m). This effect was not exacerbated by exposure to 100 ppm CO at 15,000 ft. LV weights were increased by 100 ppm CO but were not affected by 15,000 ft altitude. Moreover, the 100 ppm CO effect was not present at 15,000 ft. HT weight increased with exposure to 100 ppm and/or 15,000 ft. The effect was not exacerbated by concomitant exposure to 15,000 ft.

RV weights were significantly greater than controls (Fig. 3) in animals exposed to 500 ppm CO or 15,000 ft (4,572 m). This effect was exacerbated by exposure to 500 ppm CO at 15,000 ft. LV weights were increased by 500 ppm CO and at 15,000 ft. This altitude alone, did not affect LV weights. HT increased significantly in animals

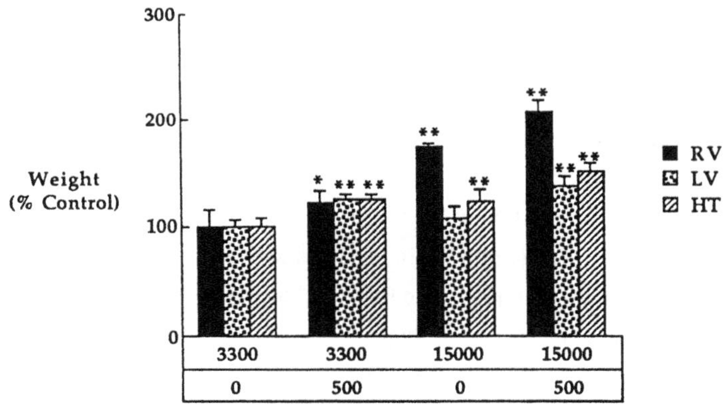

Fig. 3. Heart weights in rats exposed to 3,300 or 15,000 ft and/or 500 ppm CO. Legend as in Fig. 1.

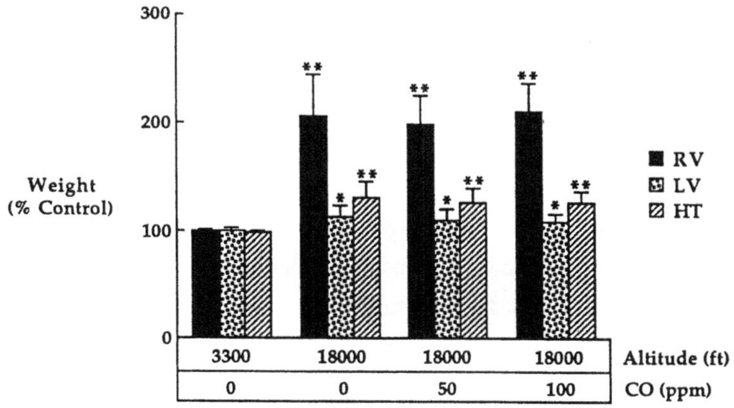

Fig. 4. Heart weights in rats exposed to 3,300 or 18,000 ft and 0, 50, 100 ppm CO. Legend as in Fig. 1.

exposed to 500 ppm CO or 15,000 ft. CO (500 ppm) at 15,000 ft further exacerbated HT weights.

RV, LV and HT weights were significantly greater than controls (Fig. 4) in animals exposed to CO (0,50,100 ppm) and 18,000 ft (5,486 m). The response to 18,000 ft appears to be complete because 50 and 100 ppm CO did not increase the RV, LV, or HT weights further. At 18,000 ft plus 500 ppm CO (not shown) most of the animals died.

Discussion

CO and altitude both produce a tissue hypoxia but the mechanisms of action are quite different. CO, by binding with hemoglobin to form COHb, reduces the amount of functional hemoglobin and makes the hemoglobin-bound oxygen less available to the tissues; this results in anemic hypoxia. On the other hand, at altitude, the inspired PO_2 and the present of hemoglobin saturated with oxygen are lower than normal, and, accordingly, the amount of oxygen delivered to the tissues is less; this results in hypoxic hypoxia.

Two other features of CO metabolism may be important to this study. First, there is a basal level of CO produced by the catabolism of heme proteins which produces COHb levels of approximately 0.6% at 3,3000 ft (Table 1). At 18,000 ft, COHb levels increase to 1.9% because the reduced PO_2 favors CO in its competition with O_2 for binding sites on the hemoglobin molecule. This effect is small in relation to the levels of COHb at which changes in the heart were observed, and is probably of little consequence. Second, there is growing evidence that CO in high concentrations, may affect calcium transport at the cell level (6,7). This effect has only been observed at extremely high CO concentrations, but is not known if chronic exposure to CO might elicit changes in calcium movement at lower CO concentrations.

In our studies, RV weights, while unaffected by 35 ppm CO, were increased significantly by 100 ppm CO and 10,000 ft altitude. Thus, the threshold for this response appears to be at 100 ppm CO or less. The right ventricular hypertrophy was exacerbated by concurrent exposure to 500 ppm CO and 15,000 ft elevation. Right ventricular hypertrophy is a prominent feature of extended stay at altitude (8,9,10). It occurs secondarily to pulmonary hypertension which imposes an increased workload on the right ventricle. Pulmonary hypertension results from the pulmonary vasoconstriction which occurs at altitude and which is mediated by the direct effects of hypoxia on arterial smooth muscle. Some degree of pulmonary vasoconstriction in response to hypoxia may be beneficial, in that it can match perfusion to ventilation within the lung (11). The response is a useful physiological mechanism in local disease of the lung (e.g., pneumonia), because the hypoxic vasoconstrictor mechanism can shunt blood away from diseased areas. But at altitude, all alveoli are hypoxic, and the vasoconstrictor response is general throughout the lung and, if continued, will produce right ventricular hypertrophy.

The well documented physiological responses to chronic CO exposure include increased oxygen-carrying capacity of the blood and cardiac enlargement. The levels of CO reported to elicit these changes are generally high, and there is some question if the changes occurring in the heart can be considered beneficial. In the studies reported here, the CO exposures were more comparable to those encountered in the ambient and occupational environment and by smokers.

LV weights were unaffected by altitudes up to 18,000 ft but were increased significantly by 100 ppm CO. The left ventricular hypertrophy was intensified by concurrent exposure to 500 ppm CO and 15,000 ft. Left ventricular hypertrophy in response to chronic CO exposure is caused by a volume overload of the left ventricle. Penney and co-workers have shown cardiac enlargement in rats exposed to 500 ppm CO for up to 42 d (12,13). At 500 ppm CO, both the right and left ventricles were enlarged and cardiac output was increased by an increased stroke volume. These workers concluded that the greater heart work required to provide adequate oxygenation of the tissues is the major factor responsible for the cardiomegaly induced by CO, and that CO produces a volume overload rather than a pressure overload on the heart.

While unaffected by 35 ppm CO and 10,000 ft altitude, HT weights were increased significantly by 100 ppm and 15,000 ft altitude. The cardiomegaly was intensified by concurrent exposure to 500 ppm CO and 15,000 ft altitude. Cardiac enlargement in response to chronic CO exposure has been reported (14). In rats exposed to 500 ppm CO, heart mass increased rapidly and cardiac lactate dehydrogenase composition changed in parallel with the heart weight changes (15). Penney and coworkers (16) concluded that the threshold for cardiac enlargement is near 200 ppm, and, unlike cardiac hypertrophy caused by altitude which involves primarily the right ventricle, cardiac hypertrophy caused by CO involves the whole heart. Our studies would indicate the threshold for cardiac hypertrophy may be closer to 100 ppm CO.

Acknowledgements

I wish to thank Debbie Hays for reviewing and typing this manuscript. Research described in this article was supported by the Health Effects Institute, an organization that is jointly funded by the United States Environmental Protection Agency and automotive manufacturers.

References

1. McGrath JJ. J Toxicol Environ Health 1988;23(3):21-28.
2. McGrath JJ. Health Effects Institute Research Report number 1989;27.
3. Small KA, Radford EP, Frazier JM, Rodkey FC, Collison HA. J Appl Physiol 1971;31:154-160.
4. Beeckmans JM. Br J Ind Med 1967;24:71-72.
5. Rodkey FL. Ann NY Acad Sci 1970;174:261-267.
6. Lin H, McGrath JJ. Life Sciences 1988;43:1813-1816.
7. Ramos K, Lin H, McGrath JJ. Biochem Pharm 1989;38:1368-1370.
8. Grover RF, Reeves JT, Will DH, Blount SG Jr. J Appl Physiol 1963:18:567-574.
9. Reeves JT, Grover EB, Grover RF. J Appl Physiol 1963;18:575-579.
10. Burton RR, Besch EL, Smith AH. Am J Physiol 1968;214:1438-1442.
11. Reeves JT, Wagner WW Jr., McMurtry F, Grover RF. In: Environmental Physiology III (ed.) RD Shaw. University Park Press, Baltimore MD 1979; pp. 289-310.
12. Penney DG, Sodt PC, Cutilletta A. Toxicol Appl Pharmacol 1979;50:213-218.
13. Penney DG, Baylerian MS, Thill JE, Fanning CM, Yedavally S. Am J Physiol 1982;243:H328-H339.
14. Theodore J, O'Donnell RD, Black KC. JOM 1971;13:242-255.
15. Penney DG, Dunham E, Benjamin M. Toxicol Appl Pharmacol 1974b;28:493-497.
16. Penney DG, Benjamin M, Dunham E. J Appl Physiol 1974a;37:80-84.

E. CARDIAC HYPERTROPHY AND FAILURE

Thyroid Hormones and Cardiac Hypertrophy

H.-G. Zimmer

Department of Physiology, University of Munich,
8000 München 2, GERMANY

Introduction

Thyroid hormones play an important role in many biological functions. They are involved in growth and development, in stimulation of metabolism and thermogenesis, and they have a permissive role for the actions of other hormones (Fig. 1). The effects of hyper- and hypo-thyroidism have marked clinical symptoms, in particular in relation to the cardiovascular system. In this contribution, the cardiac and vascular effects of triiodothyronine (T_3) as well as the development of T_3-induced cardiac hypertrophy will be discussed.

Materials and Methods

All experiments were done on female Sprague-Dawley rats (200-250 g body weight) fed a diet of Altromin^R with free access to tap water. 3,3',5-Triiodo-L-thyronine (T_3) was obtained from Sigma Chemie München and was injected daily (0.2 mg/kg, s.c.) for 3 days in animals that received continuous i.v. infusion of 0.9% NaCl or the β_1-receptor blocker metoprolol (1 mg/kg/h). The infusion rate was 1 ml/h (1). To measure cardiac functional parameters in the closed-chest, anesthetized rats, methods for catheterization of the left and right heart were developed using Millar ultraminiature catheter pressure transducers (2,3). Cardiac output was determined with the thermodilution technique (4). RNA and DNA concentrations were

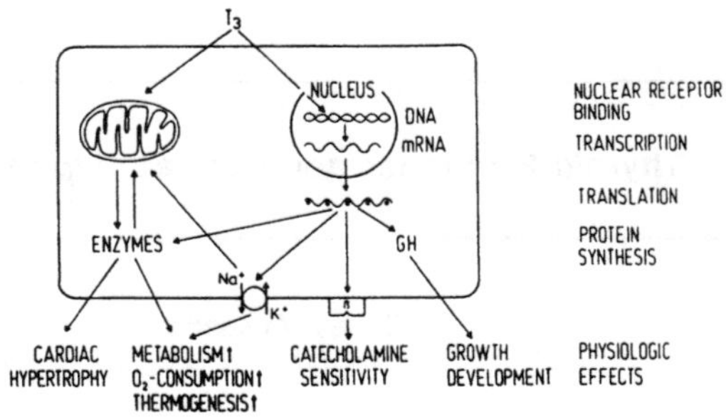

Fig. 1. Action of T_3 at the cellular level and the physiological effects.

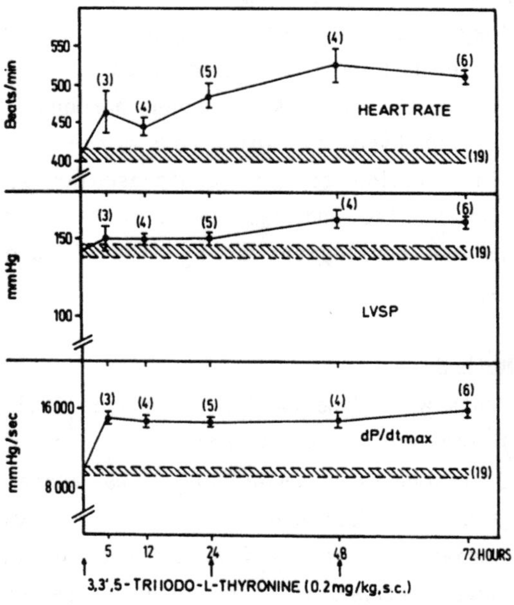

Fig. 2. T_3-induced changes in heart rate, left ventricular systolic pressure (LVSP) and LV dp/dt_{max} as measured with the Millar ultraminiature catheter pressure transducer model PR-249. Data are mean values ± SEM, number of experiments in parentheses.

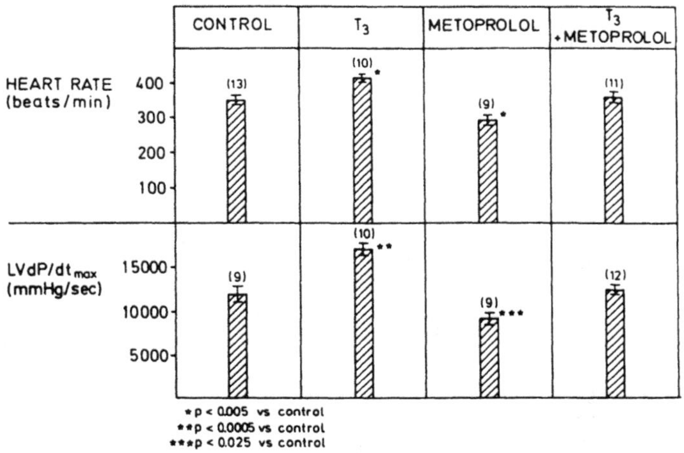

Fig. 3. Effects of T_3 and metoprolol alone and in combination on heart rate and LV dp/dt$_{max}$ afer 3 days. Mean values ± SEM, number of experiments in parentheses.

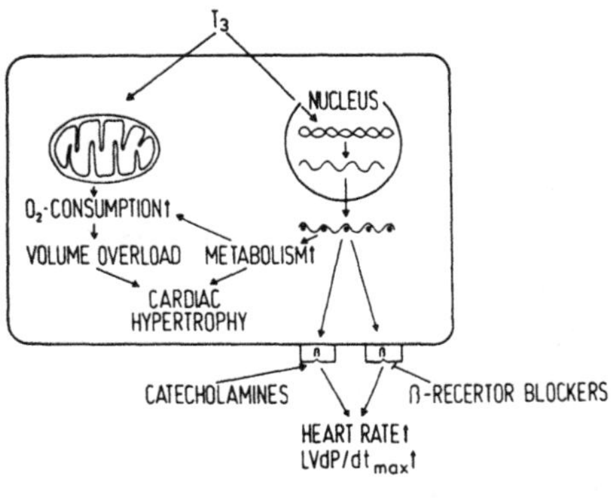

Fig. 4. Schematic presentation of the action of T_3 at the cellular level and the physiological effect on heart function and the development of cardiac hypertrophy. Increased sensitivity to catecholamines and the effect of ß-receptor blockers.

determined spectrophotometrically according to the Schmidt-Thannhauser procedure as modified by Fleck and Munro (5). Adenine nucleotide biosynthesis was measured using 1-^{14}C-glycine as precursor substance (6). Protein synthesis was also measured with 1-^{14}C-glycine (7).

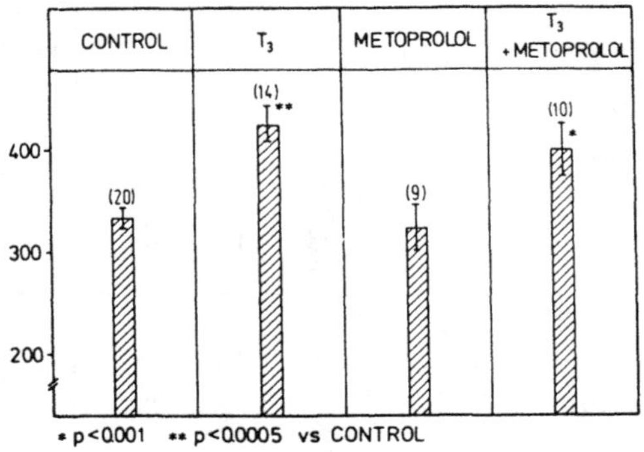

Fig. 5. Effects of T_3 and metoprolol alone and in combination on cardiac output in rats after 3 days of treatment. The data are mean values ± SEM. Number of experiments are in parentheses. Cardiac output is expressed as ml/kg/min.

Results and Discussion

When T_3 was administered, heart rate increased markedly, and LV dp/dt_{max} was about twice as high as in the control. LVSP was only moderately elevated (Fig. 2). To decide whether this positive chronotropic and intotropic effect was the exclusive result of the action of T_3 or whether catecholamines may also be involved, the β_1-receptor blocker metoprolol was administered as continuous i.v. infusion in T_3-treated rats for 3 days. The data in Fig. 3 depict the changes in heart rate and LV dp/dt_{max} in control rats and 3 days after daily administration of T_3 and continuous i.v. infusion of metoprolol, as well as 3 days of metoprolol infusion in T_3-treated rats. T_3 induced an increase and metoprolol a decrease of both parameters. When metoprolol was applied in T_3-treated animals, the T_3-induced positive chronotropic and inotropic effects were entirely prevented.

This effect can easily be explained on the basis of radioligand studies which have revealed that the number of ß-adrenergic receptors in cardiac membranes from hyperthyroid animals is increased roughly twofold (8). This leads to an increased sensitivity to catecholamines. As a consequence, the positive chronotropic and inotropic effects can be abolished by ß-receptor blockers (Fig. 4).

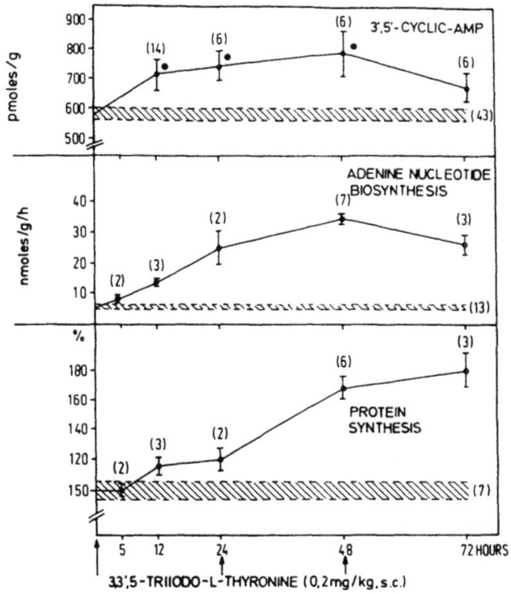

Fig. 6. T_3-induced changes in cardiac cyclic AMP, adenine nucleotide and protein synthesis. Mean values ± SEM, number of experiments in parentheses.

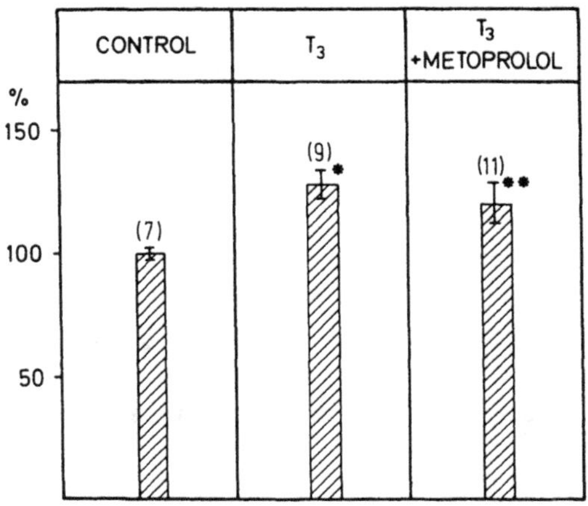

Fig. 7. Effect of metoprolol on the T_3-elicited increase in cardiac cyclic AMP. Mean values ± SEM, number of experiments in parentheses.

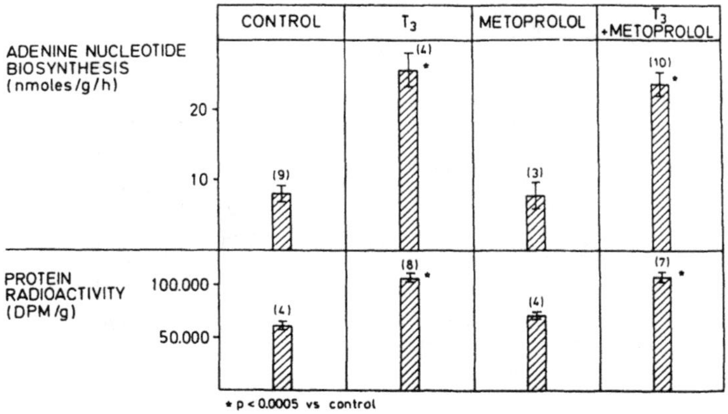

Fig. 8. Changes in cardiac adenine nucleotide biosynthesis and in the radioactivity of myocardial proteins due to incorporation of 1-^{14}C-glycine in rats after 3 days of treatment with T_3 and with metoprolol alone and in combination. Data are mean values ± SEM, number of experiments in parentheses.

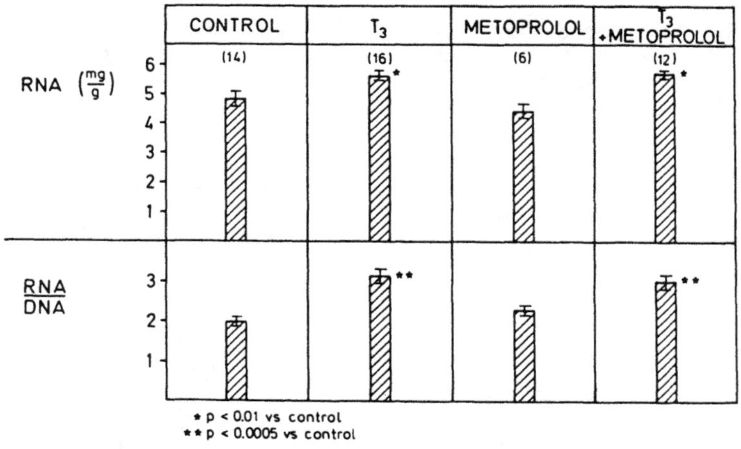

Fig. 9. RNA concentration and RNA/DNA ratio in hearts of rats under the influence of T_3 and metoprolol alone and in combination. Mean values ± SEM, number of experiments in parentheses.

But what about the increase in O_2-consumption and in metabolism that occurs under the influence of T_3? Can this also be affected by ß-receptor blockade? To decide this, cardiac output and several

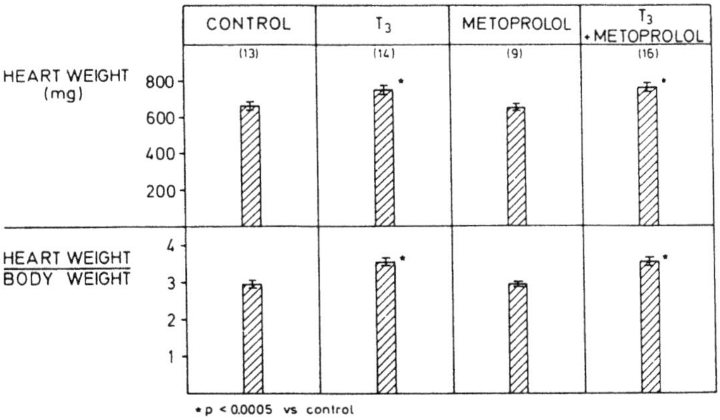

Fig. 10. Heart weight and heart weight/body weight ratio of rats treated with T_3 or with metoprolol alone and in combination for 3 days. Data are mean values ± SEM, number of experiments in parentheses.

metabolic parameters were measured. T_3 induced a marked increase in cardiac output, metoprolol had no appreciable influence and did not significantly affect the T_3-induced increase. Thus, cardiac output was as high as in the group of animals treated with T_3 alone (Fig. 5). Since the T_3-elicited increase in heart rate was normalized with metoprolol (Fig. 3), the increase in cardiac output must have been maintained by an increase in stroke volume. Thus, volume overload of the heart persisted despite effective ß-receptor blockade (Fig. 4).

It was now of interest to examine whether the T_3-induced increase in the metabolic parameters was affected by ß-receptor blockade. T_3 induced an increase in cardiac cyclic AMP, in the synthesis of myocardial adenine nucleotides and in protein synthesis (Fig. 6). When the ß-receptor blocker metoprolol was administered as continuous i.v. infusion, the increase in cAMP was not affected significantly (Fig. 7). Similar results were obtained in previous studies of other authors. In these, thyroxine elevated adenylate cyclase activity which was not affected by propranolol. Propranolol, however, prevented entirely the norepinephrine-induced increase in

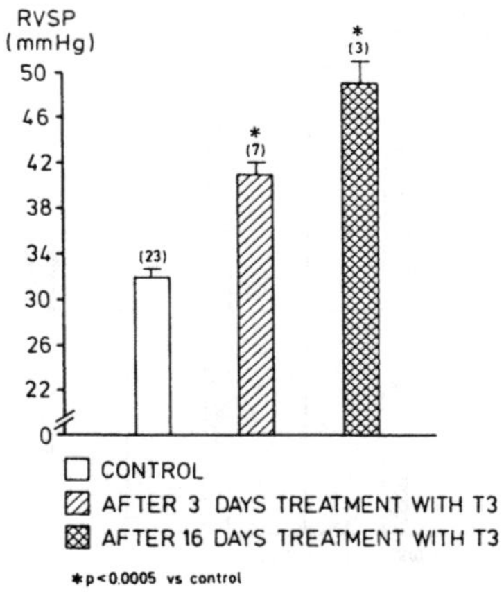

Fig. 11. Time-dependent increase in right ventricle systolic pressure (RVSP) in rats that had received daily administration of T_3. Data are mean values ± SEM, number of experiments in parentheses.

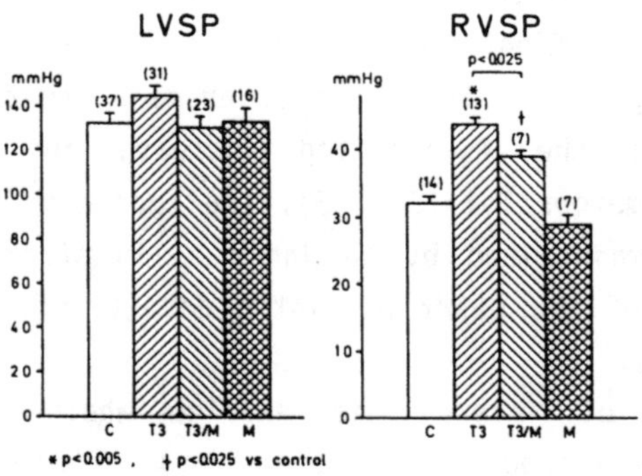

Fig. 12. Changes in left (LVSP) and right ventricular systolic pressure (RVSP) after 3 days of daily administration of T_3. C: Control; M: Metoprolol. Data represent mean values ± SEM, number of experiments are in parentheses.

adenylate cyclase activity (9). Likewise, the T_3-induced stimulation of cardiac adenine nucleotide and protein synthesis was not affected by concomitant administration of metoprolol. Metoprolol itself had no effect of its own (Fig. 8). The same was true for cardiac RNA concentration and for the RNA/DNA ratio. Both were elevated to about the same extent in the T_3-treated animals irrespective of whether or not the ß-receptor blocker was given (Fig. 9.) As a consequence, both heart weight and the heart weight/body weight ratio which were elevated by T_3 were not affected by ß-receptor blockade (Fig. 10). Thus, catecholamines seem not to play a role in the T_3-induced development of cardiac hypertrophy (10).

On the basis of our data, it cannot be decided which role volume overload on the one hand and stimulation of cardiac metabolism on the other hand, plays in the initiation and maintenance of cardiac hypertrophy (Fig. 4). Recent studies have suggested that volume overload may be decisive. In these experiments, hearts from hypothyroid donor rats were transplanted into the abdominal cavity of hypothyroid recipient rats. Administration of thyroid hormone induced a comparable increase in heart rate in the isoenzyme shift of myosin from V_3 to V_1 in both the recipient and the transplanted heart. However, the transplanted heart, in contrast to the recipient heart, did not develop cardiac hypertrophy, but underwent atrophy (11). Thus, thyroid hormone alone seems not to be a sufficient stimulus, but rather the increase in volume load.

All results that have been demonstrated so far were related to the left ventricle. Thyroid hormones, however, also affect the function and morphology of the right heart. Fig. 11 shows that there was a time-dependent elevation in RVSP after 3 and 16 days of daily T_3 injections. Also, mean pulmonary artery pressure was increased from 18 to 26 mm Hg (12). Pulmonary vascular resistance was not altered. This is due to the fact that cardiac output was increased. It can thus be concluded that RVSP and pulmonary artery pressure were elevated as a consequence of the increase in cardiac output. The right ventricle has to handle a pressure overload in addition to the volume overload. This becomes evident when the functional changes of the left ventricle are compared with those of the right ventricle. Fig. 12 compares the changes in LVSP and in RVSP after 3 days of daily T_3-treatment alone and in combination with metoprolol. There was only a slight increase in LVSP by 10% which was prevented by ß-receptor blockade. RVSP, however, was much more elevated (+ 34%).

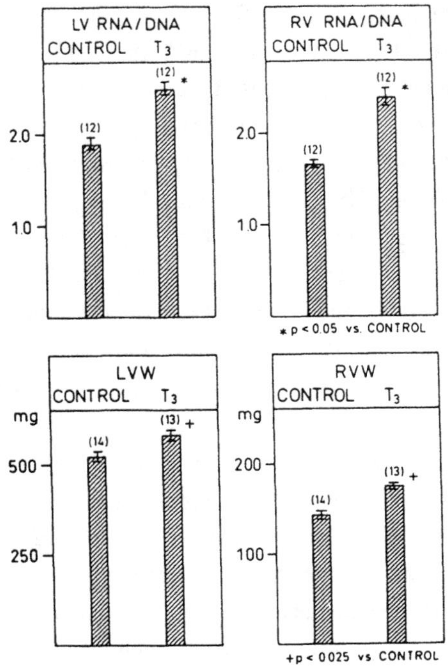

Fig. 13. Changes in the left ventricular (LV) and right ventricular (RV) RNA/DNA ratio and in LV and RV weight after 3 days of T_3.

This elevation was attenuated, but not abolished by metoprolol. Metoprolol alone had no appreciable effect. Thus, there was a disproportionate increase in RVSP, and this could not be prevented by ß-receptor blockade (12).

Based on these asymmetric changes in left and right heart function, one could expect that also the development of cardiac hypertrophy may be disproportionate. Cardiac hypertrophy was measured in terms of a metabolic and morphologic parameter. After 3 days of daily T_3-treatment, the RNA/DNA ratio was increased both in the left and right heart. However, the elevation in the right heart was 45% compared to 32% in the left heart. This was also reflected in the different changes in ventricular weight. Left ventricular weight was increased by 10%, right ventricular weight by 22% (Fig. 13).

Thus, T_3 has different effects on the peripheral and pulmonary circulation. Total peripheral resistance is lower, but pulmonary vascular resistance is unchanged. Since cardiac output is increased, there is pressure overload of the right ventricle in addition to volume overload. These different functional changes are exactly reflected in the disproportionate development of cardiac hypertrophy which affects the right ventricle to a greater extent than the left ventricle.

In summary, thyroid hormones have characteristic effects in relation to the heart:

1. Thyroid hormones bind to nuclear receptors, induce a change in gene expression and lead to increased synthesis of various proteins. Among these, the shift in myosin isoenzyme pattern plays an important role for the increase in contractile performance of the heart.

2. The positive chronotropic and inotropic effects of T_3 are mediated by catecholamines.

3. The T_3-induced cardiac hypertrophy seems not to be exclusively the result of metabolic stimulation, but to an appreciable extent the consequence of the increased volume overload.

4. The T_3-induced right ventricular hypertrophy is more pronounced than left ventricular hypertrophy. This disproportionate development of cardiac hypertrophy reflects pressure overload of the right ventricle in addition to volume overload.

References

1. Zimmer H-G, Ibel H. Experientia 1979;35:510-511.
2. Zimmer H-G. Basic Res Cardiol 1983;78:77-84.
3. Zimmer H-G, Zierhut W, Seesko RC, Verekamp AE. Basic Res Cardiol 1988,83:48-57.
4. Zimmer H-G, Zierhut W, Marschner W. J Mol Cell Cardiol 1987;19:635-639.
5. Fleck A, Munro HN. Biochim Biophys Acta 1962;55:571-583.
6. Zimmer H-G, Trendelenburg C, Kammermeier H, Gerlach E. Circ Res 1973;27:635-642.
7. Zimmer H-G, Steinkopff G, Gerlach E. Pflügers Arch 1972;336:311-325.

8. Williams LT, Lefkowitz RJ, Watanabe AM, Hathaway DR, Besch HR. J Biol Chem 1977;252:2787-2789.
9. Levey GS, Epstein SE. J Clin Invest 1969;48:1663-1669.
10. Zierhut W, Zimmer H-G. Basic Res Cardiol 1989;84:359-370.
11. Korecky B, Zak R, Schwartz K, Aschenbrenner V. Circ Res 1987;60:824-830.
12. Zierhut W, Zimmer H-G. J Mol Cell Cardiol 1989;21:617-624.

Diastolic Dysfunction of the Heart. Pharmacological Strategies for Modulating Calcium Sequestration of the Sarcoplasmic Reticulum

H. Rupp

Institute of Physiology II, University of Tübingen, Tübingen, FEDERAL REPUBLIC OF GERMANY

Despite great research efforts, the adequate therapy of heart failure remains a challenge. With the exception of valvular disease or certain aspects of coronary heart disease which can be corrected by surgery or angioplasty, the treatment of heart failure is symptomatic. The available therapeutic regimens are typically aimed at correcting the hemodynamic abnormalities by improving the contractile performance of the heart, by facilitating oxygen supply to the heart or by unloading the heart. In severe heart failure, the neuro-humoral activation involving the renin-angiotensin system and the sympathetic nervous system are additional targets. Although the established drugs treat certain disorders of the heart failure patient, they often prove to be less efficient in increasing life expectancy or quality of life (1). It thus appears that the current drug approaches cannot normalize the major defects present in patients with a progressive deterioration of heart function.

The drugs targeted at the heart are typically designed to modulate the function of one out of a multitude of cellular components involved in the regulation of contractile performance. Such an approach does not take into account the recent findings indicating that a main problem of diseased hearts could arise from a dysregulation of gene expression involving various structural elements of the heart (2). Because the heart represents a dynamic structure determined by the equilibrium between protein synthesis and degradation (3), a deranged gene expression would necessarily give rise to qualitative or quantitative alterations of the respective proteins. Specific changes in isogene expression are currently well documented for the myosin heavy chains of the rat heart subjected to various pathophysiological loads (4,5). There is, however, increasing evidence that the gene expression of components involved in the ion homeostasis of the heart muscle cell are also affected. Because the ion equilibrium of the heart muscle cell depends on the concerted action of various structural components a deranged expression of genes would necessarily be associated with an impaired heart function. Obviously a disordered organization of the systems involved in ion homeostasis is a progressive event and it is, therefore, not surprising that heart failure is a continuum ranging from patients with normal systolic function but an impaired exercise tolerance to patients with the characteristic signs of congestive heart failure (6).

In view of the potential role of an impaired signal transduction of gene expression in the pathogenesis of heart disease, we initiated studies to identify the cellular signals affecting isogene expression in hearts subjected to various physiological and pathophysiological loads. In parallel to the myosin isogene expression, we studied biochemical and functional parameters of the sarcoplasmic reticulum (SR) Ca^{2+} pump which is the major determinant of cellular Ca^{2+} sequestration during diastole (7,8). In this overview, emphasis is placed on the regulation of the SR Ca^{2+} pump mechanism in normal and diseased hearts and interventions are described which acutely modulate the Ca^{2+} pump activity. Furthermore, a completely different approach is described which attempts to modulate signal transduction for the expression of the SR Ca^{2+} pump.

Prevalence of Diastolic Dysfunction

The significance of the diastolic phase for heart performance has only recently been fully recognized and the techniques for assessing the determinants of cardiac relaxation have been developed (9-11). The importance of the relaxation phase of heart muscle in the pharmacological management of heart failure has also been underestimated for many years. To stress the importance of cardiac relaxation, the term lusitropic has been coined for systems involved in the control of cardiac relaxation and a positive lusitropic effect of drugs has been considered as a beneficial property (12-14). Support for the clinical relevance of an impaired diastolic function is provided by studies which show that patients with congestive heart failure either have an impaired systolic and diastolic function or an adequate systolic function but exhibit signs of diastolic dysfunction (15,16). This indicates that the diastolic dysfunction can be present in mild-to-severe congestive heart failure and is not necessarily coupled to systolic failure, whereas systolic failure alone seems not to occur (17). Although diastolic dysfunction can arise also from alterations in the viscoelastic properties mainly due to fibrosis, an important component involves abnormalities in the cellular systems responsible for Ca^{2+} sequestration during diastole. The Ca^{2+} removal involves the SR Ca^{2+} pump mechanism, the sarcolemmal Na^{+}-Ca^{2+} exchanger and the sarcolemmal Ca^{2+} pump (8,18,19). Only the SR Ca^{2+} pump has currently been characterized in detail.

An important consequence of a reduced diastolic relaxation relates to the fact that the subendocardial coronary perfusion (20) is impaired. Ischemia curtails not only the energy-dependent Ca^{2+} uptake of SR but has additional detrimental effects on the SR itself (21). A reduced coronary perfusion is thus expected to induce a self-perpetuating cycle with respect to diastolic dysfunction. The reduced perfusion is expected to contribute also to the deleterious geometric remodeling of the heart leading to chamber dilatation (22). An important aspect of an impairment of Ca^{2+} sequestration relates to the increased risk of arrhythmogenesis observed in patients with diastolic dysfunction. In accordance, it was found that abnormal pacemaking of human atrial myocardium is strongly modulated by SR-dependent processes (23).

The clinical findings of a slowed relaxation in heart failure patients can also be seen in experimental animals with a well defined overload of the heart. The total contraction duration of the

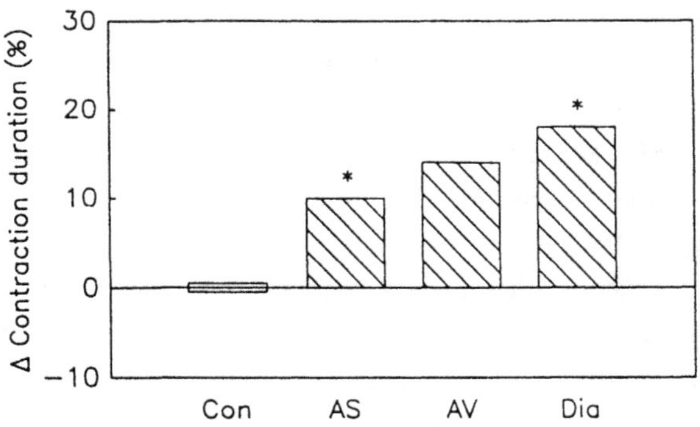

Fig. 1. Relative change in total contraction duration of isometrically contracting papillary muscles from rats with pressure overload of the heart due to stenosis of the abdominal aorta (AS), volume overload due to abdominal arteriovenous shunt (AV) and rats with streptozotocin-induced diabetes (Dia). The increase in contraction duration is given as a percentage of the contraction duration of the respective control rats; statistical comparisons were made between experimental and the respective control rats by Student's t-test. * $P < 0.05$. The data are adapted from Ref. 24.

mechanogram of isometrically contracting papillary muscles was increased in rats with a pressure overload of the heart due to aortic stenosis or with a volume overload arising from an arteriovenous shunt (Fig. 1). It is noteworthy that not only a hemodynamic overload but also the deranged cardiac metabolism arising from diabetes mellitus resulted in a prolonged twitch duration (Fig. 1). Because an increase in the total twitch duration reflects a slowed relaxation in the rat heart (25), it appears that an impaired relaxation is a characteristic phenomenon of diseased hearts with various etiologies.

Although the mechanical activity of rat papillary muscles depends not only on the rate of Ca^{2+} sequestration by the SR but is determined also by the sarcolemmal structures involved in Ca^{2+} removal from the myocyte, one would nonetheless expect that the activity of the SR Ca^{2+} pump is depressed in functional states characterized by a slowed relaxation. In accordance, in pressure overloaded hearts the SR Ca^{2+}-stimulated ATPase activity was reduced to a similar extent as in diabetic hearts (26,27).

Acute Regulation of SR Function

The major regulation of the Ca^{2+} uptake mechanism of SR arises from the cyclic AMP-dependent and Ca^{2+}-dependent phosphorylation of phospholamban (28-32) and there is clear evidence that the relaxation rate of isolated hearts is linked to the degree of phospholamban phosphorylation (33). In addition to phospholamban phosphorylation, a number of other modulatory mechanisms have been identified involving protein kinase C (34), methylation of SR phosphatidylethanolamine (35,36), insulin (37,38) and an endogenous Ca^{2+} transport inhibitor (39). Intriguing is the finding that free fatty acids can have a distinct effect on SR. Palmitic acid increased the rate of Ca^{2+} uptake of isolated SR vesicles, whereas the monounsaturated oleic acid had an inhibitory action (40-42). Long-chain acylcarnitines were reported to depress the Ca^{2+} pump activity (43) which would be of particular relevance in diabetic (44,45) and ischemic hearts (21).

The activity of the SR Ca^{2+} pump is not only important for the relaxation of the heart but is essential also for the amount of SR Ca^{2+} which can be released during the onset of systole. Compared with the Ca^{2+} pump mechanism, the SR Ca^{2+}-induced Ca^{2+} release via the SR Ca^{2+} channels (46-49) is more difficult to quantitate and its role in the pathogenesis of heart disease is currently not understood.

Agents with an Acute Effect on the SR Ca^{2+} Pump Mechanism

The characterization of the SR Ca^{2+} pump mechanism led to the discovery of a number of agents which inhibit the Ca^{2+} uptake. Quinidine (50), quercetin (51), cyclopiazonic acid (52,53), suramin (54), the alkaloid sanguinarine (55), the snake venom notexin (56) and the environmental pollutants DDT (57) and kerosene (58) had an inhibitory action. In this respect, it should be mentioned that a chronic alcohol ingestion depressed also the Ca^{2+} pump activity (59-62). An inhibitory effect was reported also for a number of drugs. Polymyxin B (63), chlorpromazine (50), trifluoroperazine (64), pentobarbital (65), the analgesic diflunisal (66), the local anaesthetics lidocaine and procaineamide (67) and the Ca^{2+} channel agonist BayK 8644 (68,69) reduced the SR Ca^{2+} uptake. Noteworthy is that also the ß-adrenergic blocking drug propranolol reduced the rate of Ca^{2+} uptake and it was shown that this effect was independent of its negative inotropic and ß-adrenergic blocking action (70-74). The ß-adrenergic blocking drug timolol had a less inhibitory action (71). It remains, however, to be shown whether these drugs affect the SR Ca^{2+} pump mechanism also at the therapeutical doses reached in patients.

Only a few agents have been characterized which stimulate the SR Ca^{2+} pump mechanism without increasing the phosphorylation of phospholamban. Of particular interest is the finding that the dihydropyridine Ca^{2+} antagonists felodipine (75,76), nisoldipine (76,77) and nitrendipine (75) increased the rate of SR Ca^{2+} uptake whereas prenylamine (75), varapamil (75,76) and diltiazem (76) had an inhibitory action. It remains again to be shown to what extent the therapeutic effect of these Ca^{2+} antagonists is governed by their modulatory influence on the SR Ca^{2+} pump mechanism in patients. A stimulatory action on SR Ca^{2+} uptake has also been observed for gingerol (78), low concentrations of the divalent cation inophore A23187 (79) and low concentrations of vanadate (80,81) which can mimic some of the effects of insulin. It should also be mentioned that the pharmacological action of diethyl ether as a muscle relaxant has been linked to the stimulation of SR Ca^{2+} uptake (82).

Agents with an Acute Effect on SR Ca^{2+} Channels

Compared with the SR Ca^{2+} pump, much less is known on the possibilities of modulating the action of the SR Ca^{2+} channels (18,83,84). Agents which block Ca^{2+} release are ruthenium red (85-87), neomycin (88,89), FLA 365 (88,89), ryanodine (87,90,91) butanedione monoxime (92) and thymol (93). It appears, however, that the interference with the SR Ca^{2+} channels provides an imbalance which is difficult to control. Thus, the anti-cancer agent doxorubicin was shown to increase the open probability of SR Ca^{2+} channels and it is thought that this may contribute to the cardiotoxicity of anthraquinones (94). An increased Ca^{2+} efflux could also be induced by chemical modification of the SR with methylbenzimidate (95). Although agents interfering with the SR Ca^{2+} release channels are of pharmacological interest, clearly more work is required to assess their therapeutic potential.

Chronic Regulation of SR Function

A completely different approach for improving diastolic function would be directed at the modulation of gene expression of SR structures involved in cellular CA^{2+} homeostasis. Currently, the sequences of the coding genes of the major sarcoplasmic reticulum proteins are known. For the Ca^{2+} pump at least five distinct isoforms have been identified: the adult fast-twitch muscle isoform and the alternatively spliced neonatal isoform (96,97), the cardiac/slow-twitch muscle isoform (98,99), its alternatively spliced smooth muscle/non-muscle isoform (100,101) and an isoform found in various tissues (102). Phospholamban does not exist in isoforms and the same protein is expressed in cardiac and slow-twitch muscle (103,104). The Ca^{2+}-release channel (ryanodine receptor) occurs in two isoforms; the cardiac isoform is expressed also in brain whereas the skeletal isoform occurs in slow- and fast-twitch skeletal muscle (105,106). The Ca^{2+} binding protein calsequestrin which is located within the lumen of the SR occurs also in two isoforms, a cardiac and a fast-twitch muscle form (107-109).

The promoter and enhancer sequences of the respective genes have, however, not been fully examined and most probably not all transcription factors have been identified. Best documented is the effect of thyroid hormones on mRNA levels of the SR Ca^{2+} pump and phospholamban. In hypothyroid rats, the mRNA levels of the Ca^{2+} pump and of phospholamban were reduced (98,110,111). In hyperthyroid rats the mRNA level of the Ca^{2+} pump was increased, however, the mRNA of phospholamban was reduced (98,110,111). Although altered mRNA levels could arise also from changes in posttranscriptional events, it seems likely that the gene expression of the Ca^{2+} pump and phospholamban does not occur in a coordinated manner. Noteworthy is that the mRNA levels of the Ca^{2+} release channel were affected to a similar degree as the mRNA of the Ca^{2+} pump (111). The mRNA level of calsequestrin was not influenced by the thyroid status

(111). In accordance, an altered thyroid influence was shown to affect the rate of SR Ca^{2+} uptake but not the amount of Ca^{2+} taken up (112). Myothermal measurements also clearly demonstrated that there are no differences between euthyroid and hyperthyroid hearts in the amount of Ca^{2+} cycled during contraction and relaxation (113,114).

The most active thyroid hormone with respect to gene regulation is triiodothyronine which binds to a specific nuclear transcription factor. This thyroid receptor regulates gene transcription after binding to a thyroid responsive element belonging to the promoter of the respective gene. There is, however, increasing evidence that the SR genes are not only regulated by triiodothyronine but also by other signals. Thus, retinoic acid also stimulated the Ca^{2+} pump gene expression by a mechanism which did, however, not involve the thyroid hormone responsive element (115).

In the hypertrophied pressure overloaded rat and rabbit hearts, the expression of the SR Ca^{2+} pump was reduced which contributes to the depressed rate of Ca^{2+} uptake and slowed relaxation (98,116-118). Also in patients with end-stage heart failure, the expression of the Ca^{2+} pump was markedly depressed (119). Less clear are currently the mechanisms resulting in the depressed relaxation of diabetic hearts because there was no indication for a reduced expression of the Ca^{2+} pump (120). Obviously, the expression of phospholamban and particularly the phosphorylation state of phospholamban need to be examined in diabetic hearts. The pathophysiological significance of phospholamban phosphorylation is clearly documented by the finding that bacterial endotoxins reduce the phosphorylation (121).

Although thyroid hormones provide a potent signal for affecting the SR Ca^{2+} pump expression, there is general agreement that circulating thyroid hormones are not altered in a number of functional states associated with an altered Ca^{2+} uptake. In rats with pressure overloaded hearts, no evidence for a reduced thyroid influence could be found (5). In diabetic rats, circulating thyroid hormones are reduced but their injection did not normalize the depressed rate of Ca^{2+} uptake (122). Although it is currently not known whether the nuclear thyroid receptors are altered in the overloaded or diabetic hearts, it appears that additional signals play a role. Thus it was shown that in addition to thyroid hormones, catecholamines affect myosin heavy chain expression (123,124). Even if a catecholamine signal had an influence on SR genes, one would not expect that in overloaded and diabetic hearts the adrenergic influence is altered in the same manner leading to the observed equidirectional changes in the ATPase activity (26,27).

Starting from the intriguing finding that the reduced SR Ca^{2+} pump activity of diabetic hearts could at least not solely be explained by thyroid hormones, the possibility of a signal related to the deranged cardiac metabolism was examined. To simulate a reduced insulin influence on heart, rats were intermittently fasted and were either provided with tap water or tap water containing 0.8% sucrose. The Ca^{2+}-stimulated ATPase activity of the SR Ca^{2+} pump was reduced in the rats receiving tap water not, however, in the rats receiving sucrose solutions (Fig. 2). The preventive action of the sucrose feeding was independent of changes in circulating thyroid hormones (26). Because an increased carbohydrate intake is known to result in an enhanced insulin release probably mediated by incretins (125), the sucrose feeding could counteract the reduced insulin release during fasting. These findings demonstrate that the Ca^{2+}-stimulated ATPase activity of SR responds to signals which are either directly related to the insulin influence or to altered cardiac metabolism. It is noteworthy that also in the pressure overloaded heart, the reduction of the SR Ca^{2+}-stimulated ATPase activity could be prevented by providing sucrose drinking solutions (Fig. 2).

Further evidence for a role of "metabolic" signals came from the effect of etomoxir which inhibits the mitochondrial carnitine palmitoyltransferase (CPT) I and, thereby, reduces fatty acid oxidation (126,127). Treatment with etomoxir increased the SR Ca^{2+} ATPase activity in normal, diabetic and pressure overloaded hearts (Fig. 2). In normal or pressure overloaded hearts, a reduction in fatty acid oxidation is expected to be compensated by an increased glycolytic flux, whereas in diabetic hearts, the reduced glucose

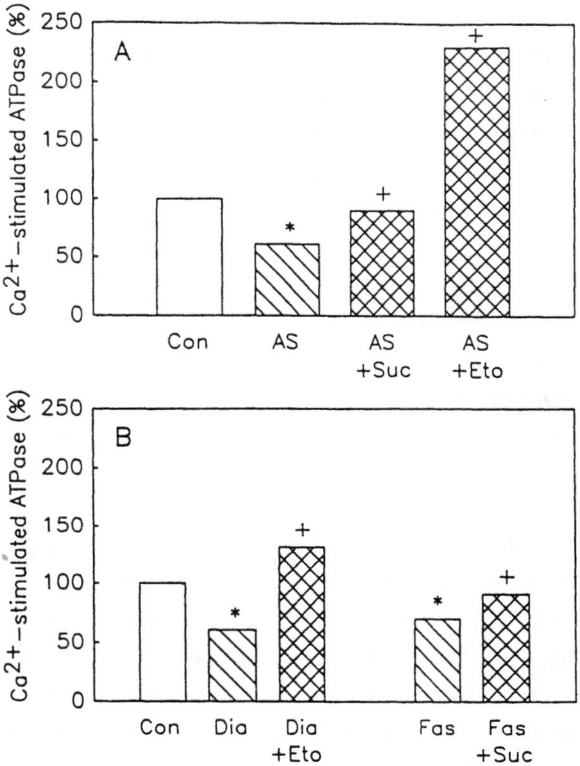

Fig. 2A. Changes in Ca^{2+}-stimulated SR ATPase activity of rats with abdominal aortic stenosis which were either untreated (AS), treated with drinking solutions containing 0.8% sucrose (AS + Suc) or treated with 15 mg/kg body wt etomoxir (AS + Eto). The changes in ATPase activity are given as a percentage of the corresponding activities of the control rats; statistical comparisons were made between untreated rats with aortic stenosis and control rats (* $P < 0.05$) or between treated and untreated rats with aortic stenosis ($^{+}$ $P < 0.05$). The data are adapted from Ref. 27 and 128.

Fig. 2B. The change in Ca^{2+}-stimulated SR ATPase activity of rats with streptozotocin-induced diabetes which were either untreated (Dia) or treated with 15 mg/kg body wt etomoxir (Dia + Eto) and intermittently fasted rats (24 h fasting followed by 24 h feeding) which were either untreated (Fas) or treated with drinking solutions containing 0.8% sucrose (Fas + Suc). The changes in ATPase activity are given as a percentage of the activity of the control rats; statistical comparisons were made between untreated rats with diabetes or intermittently fasted rats and control rats (* $P < 0.05$) or between treated and untreated rats ($^{+}$ $P < 0.05$). The data are adapted from Ref. 26 and from unpublished findings of Rupp H., Elimban V., and Dhalla, N.S.

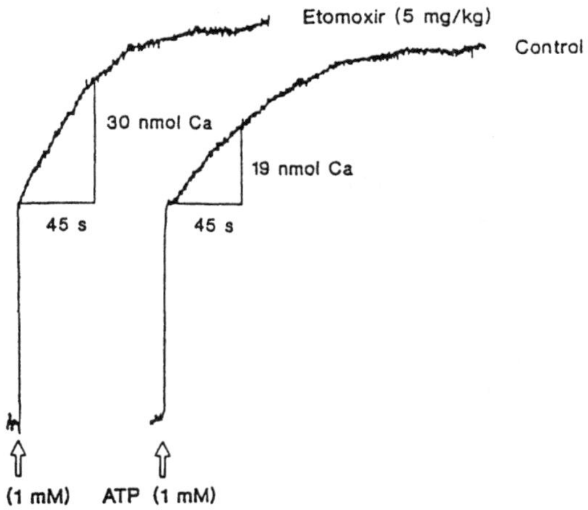

Fig. 3. Representative recordings of SR Ca^{2+} uptake of untreated rats and rats treated with 5 mg/kg body wt etomoxir. Ca^{2+} uptake was determined using arsenazo III and a dual wavelength spectrometer. The rate of Ca^{2+} uptake of control rats was 58 ± 15 nmol/mg/min and was significantly ($P < 0.05$) increased (95 ± 14 nmol/mg/min) in etomoxir-treated rats. The data are adapted from Ref. 129. Note, the data of Fig. 2 and 3 do not permit conclusions on the Ca^{2+}/ATP stoichiometry of the Ca^{2+} pump (130) because different assay conditions were used.

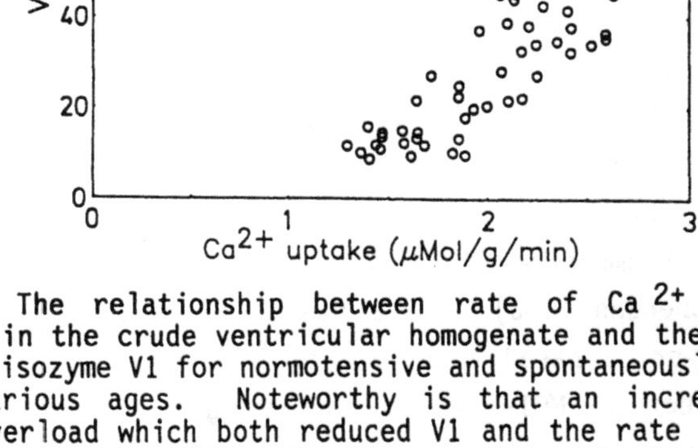

Fig. 4. The relationship between rate of Ca^{2+} uptake of SR determined in the crude ventricular homogenate and the proportion of the myosin isozyme V1 for normotensive and spontaneously hypertensive rats of various ages. Noteworthy is that an increasing age and pressure overload which both reduced V1 and the rate of Ca^{2+} uptake had no significantly different effect on this relationship.

uptake might be a limiting factor. Because etomoxir has no direct in vitro effect on the SR Ca^{2+} ATPase activity (128), it differs from all agents with an acute effect. Work is in progress to determine its effect on the gene expression of the SR Ca^{2+} pump and phospholamban. Noteworthy is that etomoxir increased in normal rats also the rate of Ca^{2+} uptake of SR vesicles (129) (Fig. 3).

It can thus be concluded that metabolic signals exist which seem to reflect the balance between fatty acid oxidation and glycolytic flux. The unraveling of these signals could provide the basis for a rational treatment of an impaired diastolic function of the heart. Of particular importance will be the identification of signals which in a coordinated manner affect multiple structures involved in the excitation-contraction process. There is indeed evidence that in a number of functional states a higher rate of Ca^{2+} sequestration is associated with an increased potential for fast contraction as is exemplified in Fig. 4. In normotensive and spontaneously hypertensive rats of various ages a relationship between rate of SR Ca^{2+} uptake and the proportion of myosin isozyme V1 was observed. Further work is required to determine whether pressure overload and increasing age reduce the rate of SR Ca^{2+} uptake and the proportion of the isozyme V1 by the same mechanisms.

Established and Novel Drugs with a Positive Lusitropic Action

The approach of modulating gene expression of systems involved in the relaxation process differs from the established drug regimens exhibiting a positive lusitropic effect. Currently best documented is a positive lusitropic effect for drugs which increase cellular cyclic AMP. Thus, the ß1-adrenergic blocking drug xamoterol with an intrinsic sympathomimetic activity was shown to have positive lusitropic effects (131-133). Also the phosphodiesterase III inhibitors (134) amrinone (135-137), milrinone (135-139), enoximone (135) and pimobendan (140) have a positive lusitropic action which requires, however, a basal adenylate cyclase activity. The positive lusitropic effect of MCI-154 was thought not to arise soley from cyclic AMP-dependent processes (141).

A negative lusitropic action was shown for DPI 201-106 which delays sodium channel inactivation (142,143). Because also the antiarrhythmic potassium channel blocker E4031 exhibited negative lusitropic effects (142), a link between prolongation of the action potential duration and a negative lusitropic action was postulated (142). Whether all agents which prolong the action potential duration have negative lusitropic effects requires further investigation. Another class with negative lusitropic action consists of agents with Ca^{2+}-sensitizing properties. Because cyclic AMP reduces the Ca^{2+} sensitivity of the contractile apparatus which contributes to the accelerated muscle relaxation, it is not unexpected that Ca^{2+}-sensitizing drugs have an opposite effect. For EMD 53998 it was indeed recently shown that the positive inotropic action was associated with an increased end-diastolic pressure (144). Based on the concept that diastolic dysfunction is an early sign of congestive heart failure, it might thus be attractive to examine the therapeutic potential of agents with a Ca^{2+}-desensitizing effect.

Taken together, the established drugs with a proven positive lusitropic action rely on the formation of increased levels of cellular cyclic AMP. Because cyclic AMP has proarrhythmogenic properties in certain pathophysiological states (145), novel positive lusitropic drugs should act independently of cellular cyclic AMP levels. Because in heart failure patients the expression of the SR Ca^{2+} pump is depressed, an upregulation of the expression of this gene could provide a novel therapeutical approach. Such a drug design is, however, closely linked to an improved understanding of the signals controlling the expression not only of the SR Ca^{2+} pump, but also of the sarcolemmal structures involved in Ca^{2+} handling of the heart muscle cell.

Acknowledgements

This research was supported by grants from the German Research Foundation (Ru 245/6-1) and the Medical Research Council of Canada (to N.S. Dhalla).

References

1. Klamerus KJ. Clin Pharmacol 1986;5:481-498.
2. Swynghedauw B. Eur Heart J 1990;11(Suppl. G):87-94.
3. Zak R. In: The Regulation of Heart Function. Basic Concepts and Clinical Applications (ed.) H Rupp. Thieme Medical Publ, New York, 1986; pp. 249-260.
4. Rupp H, Jacob R, Dhalla NS. In: The Diabetic Heart (eds.) M Nagano, NS Dhalla. Raven Press, New York, 1991; pp. 271-279.
5. Rupp H, Jacob R, Dhalla NS. In: Subcellular Basis of Contractile Failure (eds.) B Korecky, N.S. Dhalla. Kluwer Academic Publ, Boston, 1990; pp. 135-154.
6. Harris P. In: Pathophysiology and Pharmacology of Heart Disease (eds.) IS Anand, PL Wahi, NS Dhalla. Kluwer Academic Publ, Boston 1989; pp. 135-143.
7. Hasselbach W. Basic Res Cardiol 1980;75:2-12.
8. Dhalla NS, Pierce GN, Panagia V, Singal PK, Beamish RE. Basic Res Cardiol 1982;77:117-139.
9. Brutsaert DL, Housmans PR, Goethals MA. Circ Res 1980;47:637-652.
10. Brutsaert DL, Sys SU. Physiol Rev 1989;69:1228-1315.
11. Brutsaert DL, Sys SU. Circulation 1991;83:1444-1449.
12. Katz AM. J Am Coll Cardiol 1988;11:438-445.
13. Katz AM. J Am Coll Cardiol 1989;13:513-523.
14. Katz AM. Circulation 1990;82(Suppl 2):I7-I11.
15. Warnowicz MA, Parker H, Cheiltin MD. Circulation 1983;67:330-334.
16. Dougherty AH, Naccerelli GV, Gray EL, Hicks CH, Goldstein RA. Am J Cardiol 1984;54:778-782.
17. Pouleur H, Hanet C, Gurne O, Rousseau MF. Br J Clin Pharmacol 1989;28(Suppl 1):41S-52S.
18. Fleischer S, Inui M. Annu Rev Biophys Biophys Chem 1989;18:333-364.
19. Horackova M. Can J Physiol Pharmacol 1989;67:1525-1533.
20. Berne RM, Rubio R. Coronary Circulation In: Handbook of Physiology, Section 2, The Cardiovascular System. Vol I, The Heart (ed.) RM Berne. Am Physiol Society, Bethesda 1979; pp. 873-952.

21. Rapundalo ST, Briggs FN, Feher JJ. J Mol Cell Cardiol 1986;18:837-851.
22. Jacob R, Brändle M, Dierberger B, Rupp H. Basic Res Cardiol 1991;86(Suppl 1):114-130.
23. Escande D, Coraboeuf E, Planche C. J Mol Cell Cardiol 1987;19:231-241.
24. Rupp H, Takeda N. In: Catecholamines and Heart Disesase (ed.) PK Ganguly. CRC Press, Boca Raton, 1991; pp. 217-229.
25. Mattiazzi A, Garay A, Cingolani HE. J Mol Cell Cardiol 1986;18:749-758.
26. Rupp H, Elimban V, Dhalla NS. Biochem Biophys Res Commun 1989;164:319-325.
27. Rupp H, Elimban V, Dhalla NS. Biochem Biophys Res Commun 1988;156:917-923.
28. Fedelesova M, Ziegelhöffer A. Experientia 1975;31:516-518.
29. Tada M, Kirchberger MA, Katz AM. J Biol Chem 1975;250:2640-2647.
30. Will H, Blanck J, Smettan G, Wollenberger A. Biochim Biophys Acta 1976;449:295-303.
31. Tada M, Inui M. J Mol Cell Cardiol 1983;15:565-575.
32. Ambudkar IS, Fanfarillo DT, Shamoo AE. Membr Biochem 1986;6:327-346.
33. Vittone L, Mundina C, Chiappe de Cingolani G, Mattiazzi A. Acta Physiol Pharmacol Latinoam 1988;38:213-227.
34. Rogers TB, Gaa ST, Massey C, Dosemeci A. J Biol Chem 1990;265:4302-4308.
35. Ganguly PK, Panagia V, Okumura K, Dhalla NS. Biochem Biophys Res Commun 1985;130:472-478.
36. Meij JTA, Panagia V. In: Catecholamines and Heart Disease (ed.) PK Ganguly. CRC Press, Boca Raton 1991; pp. 245-266.
37. Pierce GN, Ganguly PK, Dzurba A, Dhalla NS. Adv Myocardiol 1985;6:113-125.
38. Gupta MP, Lee SL, Dhalla NS. J Pharmacol Exp Ther 1989;249:623-630.
39. Narayanan N, Bedard P, Waraich TS. Can J Physiol Pharmacol 1989;67:999-1006.

40. Watras J, Messineo FC, Herbette LG. J Biol Chem 1984;259:1319-1324.
41. Herbette LG, Favreau C, Segalman K, Napolitano CA, Watras J. J Biol Chem 1984;259:1325-1335.
42. Rathier M, Chen LY, Messineo FC, Katz AM. Life Sci 1986;38:1733-1739.
43. Pitts BJ, Tate CA, van Winkle WB, Wood JM, Entman ML. Life Sci 1978;23:391-401.
44. Dhalla NS, Kolar F, Shah KR, Ferrari R. Cardiovasc Drugs Ther 1991;5(Suppl 1):25-30.
45. Lopaschuk GD, Tahiliani AG, Vadlamudi RV, Katz S, McNeill JH. Am J Physiol Pharmacol 1983;245:H969-H976.
46. Fabiato A. Am J Physiol 1983;245:C1-C14.
47. Lai FA, Meissner GJ. Bioenerg Biomembr 1989;21:227-246.
48. MacLennan DH. Biophys J 1990;58:1355-1365.
49. Ungeheuer M, Mizgala A, Hasselbach W. Z Naturforsch (C) 1986;41:647-651.
50. Pang DC, Briggs FN. Recent Adv Stud Cardiac Struc Metab 1975;7:421-424.
51. Shoshan V, Campbell KP, Maclennan DH, Frodis W, Britt BA. Proc Natl Acad Sci USA 1980;77:2235-4438.
52. Goeger DE, Riley RT, Dorner JW, Cole RJ. Biochem Pharmacol 1988;37:978-981.
53. Goeger DE, Riley RT. Biochem Pharmacol 1989;38:3995-4003.
54. Layton D, Azzi A. Biochem Biophys Res Commun 1974;59:322-325.
55. Faddeeva MD, Beliaeva TN. Tsitologiia 1988;30:685-690.
56. Helmke S, Howard BD. Membr Biochem 1986;6:239-253.
57. Elder JH, Morre DJ, Yunghans WN. Eur J Cell Biol 1979;19:231-238.
58. Garcia M, Gonzalez R. Toxicol Lett 1985;28:59-64.
59. Bing RJ. Circulation 1978;58:965-970.
60. Bing RJ. Fed Proc 1982;41:2443-2446.
61. Katz AM. Fed Proc 1982;41:2456-2459.
62. Ohnishi ST, Waring AJ, Fang SR, Horiuchi K, Ohnishi T. Membr Biochem 1985;6:49-63.

63. Ktenas TB, Sotiroudis TG, Evangelopoulos AE. Biosci Rep 1989;9:573-578.
64. Ho MM, Scales DJ, Inesi G. Biochim Biophys Acta 1983;730:64-70.
65. Fernandez-Salguero P, Henao F, Laynez J, Gutierrez-Merino C. Biochim Biophys Acta 1990;1022:33-40.
66. Holguin JA. Biochem. Pharmacol 1988;37:4035-4040.
67. Katz AM, Repke DI, Corkedale S, Schwarz J. Cardiovasc Res 1975;9:764-769.
68. Hryshko LV, Kobayashi T, Bose D. Am J Physiol 1989;257:H407-H414.
69. Zorzato F, Volpe P, Salviati G, Margreth A. FEBS Lett 1985;186:255-258.
70. Pang DC, Briggs FN. Biochem Pharmacol 1973;22:1301-1308.
71. Messineo FC, Katz AM. J Cardiovasc Pharmacol 1979;1:449-459.
72. Noack E, Kurzmack M, Verjovski-Almeida S, Inesi G. J Pharmacol Exp Ther 1978;206:281-288.
73. Su JY, Malencik DA. Naunyn Schmiedebergs Arch Pharmacol 1985;331:194-201.
74. Pedroni P, Mattiazzi A, Gende OA, Cingolani HE. Acta Physiol Pharmacol Latinoam 1987;37:503-519.
75. Movsesian MA, Ambudkar IS, Adelstein RS, Shamoo AE. Biochem Pharmacol 1985;34:195-201.
76. Wang T, Tsai LI, Schwartz A. Eur J Pharmacol 1984;100:253-261.
77. Sassen LM, Bezstarosti K, Verdouw PD, Lamers JM. Biochem Pharmacol 1991;41:43-51.
78. Kobayashi M, Ishida Y, Shoji N, Ohizumi Y. J Pharmacol Exp Ther 1988;246:667-673.
79. Murray JJ, Kuzmin AV, Reed PW, Levitsky DO. Am J Physiol 1985;249:H1211-H1215.
80. Hagenmeyer A, Wierichs R, Bader H. Basic Res Cardiol 1980;75:452-454.
81. Medda P, Hasselbach W. Z Naturforsch (C) 1985;40:876-879.
82. Salama G, Scarpa A. J Biol Chem 1980;255:6525-6528.
83. Wyskovsky W, Suko J. Prog Clin Biol Res 1988;252:165-169.

84. Palade P, Dettbarn C, Brunder D, Stein P, Hals GJ. Bioenerg Biomembr 1989;21:295-320.
85. Hoste AM, Sys SU, de Clerck NM, Brutsaert DL. Pflügers Arch 1988;411:558-563.
86. Gupta MP, Dixon IMC, Zhao D, Dhalla NS. Can J Cardiol 1989;5:55-63.
87. Feher JJ, Manson NH, Poland JL. Arch Biochem Biophys 1988;265:171-182.
88. Chiesi M, Schwaller R, Calviello G. Biochem Biophys Res Commun 1988;154:1-8.
89. Calviello G, Chiesi M. Biochemistry 1989;28:1301-1306.
90. Feher JJ, Lipford GB. Biochim Biophys Acta 1985;813:77-86.
91. Hasselbach W, Mizgala A. Z Naturforsch (C) 1988;43:140-148.
92. Fryer MW, Gage PW, Neering IR, Dulhunty AF, Lamb GD. Pflügers Arch 1988;411:76-79.
93. Takishima K, Setaka M, Shimizu H. J Biochem (Tokyo) 1979;86:347-353.
94. Holmberg SR, Williams AJ. Circ Res 1990;67:272-283.
95. Shoshan-Barmatz V. Biochem J 1987;243:165-173.
96. Brandl CJ, deLeon S, Martin DR, MacLennan DH. J Biol Chem 1987;262:3768-3774.
97. Brandl CJ, Green NM, Korczak B, MacLennan DH. Cell 1986;44:597-607.
98. Nagai R, Zarain-Herzberg A, Brandl CJ, Fujii J, Tada M, MacLennan DH, Alpert NR, Periasamy M. Proc Natl Acad Sci USA 1989;86:2966-2970.
99. MacLennan DH, Brandl CJ, Korczak B, Green NM. Amino-acid sequence of a Ca^{2+} + Mg^{2+}-dependent ATPase from rabbit muscle sarcoplasmic reticulum, deduced from its complementary DNA sequence. Nature 1985;316:696-700.
100. Lytton J, MacLennan DH. J Biol Chem 1988;263:15024-15031.
101. Lytton J, Zarain-Herzberg A, Periasamy M, MacLennan DH. J Biol Chem 1989;264:7059-7065.
102. Burk SE, Lytton J, MacLennan DH, Shull GE. J Biol Chem 1989;264:18561-18568.

103. Fujii J, Ueno A, Kitano K, Tanaka S, Kadoma M, Tada M. J Clin Invest 1987;79:301-304.
104. Fujii J, Lytton J, Tada M, MacLennan DH. FEBS Lett 1988;227:51-55.
105. Otsu K, Willard HF, Khanna VK, Zorzato F, Green NM, MacLennan DH. J Biol Chem 1990;265:13472-13483.
106. Imagawa T, Takasago T, Shigekawa M. J Biochem (Tokyo) 1989;106:342-348.
107. Fliegel L, Ohnishi M, Carpenter MC, Khanna VK, Reithmeier RAF, MacLennan DH. Proc Natl Acad Sci USA 1987;84:1167-1171.
108. Scott BT, Simmerman HKB, Collins JH, Nadal-Ginard B, Jones LR. J Biol Chem 1988;263:8958-8964.
109. Zarain-Herzberg A, Fliegel L, MacLennan DH. J Biol Chem 1988;263:4807-4812.
110. Rohrer D, Dillmann WH. J Biol Chem 1988;263:6941-6944.
111. Arai M, Otsu K, MacLennan DH, Alpert NR, Periasamy M. Circ Res 1991;69:266-276.
112. Suko J. Biochim Biophys Acta 1971;252:324-327.
113. Alpert NR, Mulieri LA. In: The Regulation of Heart Function. Basic Concepts and Clinical Applications (ed.) H. Rupp. Thieme Medical Publ, New York, 1986; pp. 292-304.
114. Blanchard EM, Mulieri LA, Alpert NR. Basic Res Cardiol 1987;82(Suppl 2);127-135.
115. Rohrer DK, Hartong R, Dillmann WH. J Biol Chem 1991;266:8638-8646.
116. Alpert NR, Blanchard EM, Mulieri LA, Nagai R, Zarain-Herzberg A, Periasamy M. Basic Res Cardiol 1989;84(Suppl 1):55-66.
117. Komuro I, Kurabayashi M, Shibazaki Y, Takaku F, Yazaki Y. J Clin Invest 1989;83:1102-1108.
118. de la Bastie D, Levitsky D, Rappaport L, Mercadier JJ, Marotte F, Wisnewsky C, Brovkovich V, Schwartz K, Lompre AM. Circ Res 1990;66:554-564.
119. Mercadier JJ, Lompre AM, Duc P, Boheler KR, Fraysse JB, Wisnewsky C, Allen PD, Komajda M, Schwartz K. J Clin Invest 1990;85:305-309.

120. Dillmann WH. In: The Diabetic Heart (ed.) M Nagano, NS Dhalla. Raven Press, New York, 1991;pp.263-270.
121. Mohammed FI, Liu MS. J Mol Cell Cardiol 1990;22:587-598.
122. Ganguly PK, Pierce GN, Dhalla KS, Dhalla NS. Am J Physiol 1983;244:E528-E535.
123. Gupta MP, Gupta M, Stewart A, Zak R. Biochem Biophys Res Commun 1991;174:1196-1203.
124. Rupp H, Berger H-J, Pfeifer A, Werdan K. Circ Res 1991;68:1164-1173.
125. Johnson DG, Bressler R. In: Diabetes Mellitus. Theory and Practice (eds.) H Rifkin, D Porte. Elsevier, New York, 1990;pp.887-895.
126. Eistetter K, Wolf HPO. Drugs of the Future. 1986;11:1034-1036.
127. Selby PL, Sherratt HSA. Trends in Pharmacol. 1989;10:495-500.
128. Rupp H, Elimban V, Dhalla NS. FASEB J 1992;6:2349-2353.
129. Rupp H, Wahl R, Hansen M. J Appl Physiol 1992;72:352-360.
130. Hasselbach W, Mizgala A. Z Naturforsch (C) 1985;40:571-575.
131. Nuttall A, Snow HM. Br J Pharmacol 1982;77:381-388.
132. Böhm M, Mittmann C, Schwinger RH, Erdmann E. Am Heart J 1990;120:1381-1392.
133. Heng MK. Clin Cardiol 1990;13:171-176.
134. Honerjäger P. Eur Heart J 1989;10(Suppl C):25-31.
135. Schlepper M, Thormann J, Kremer P, Mitrovic V, Kramer W. J Cardiovasc Pharmacol 1989;14(Suppl 1):S9-S19.
136. Honerjäger P. Am Heart J 1991;121:1939-1944.
137. DiBianco R. Am Heart J 1991;121:1871-1875.
138. Farah AE, Frangakis CJ. Basc Res Cardiol 1989;84(Suppl 1):85-103.
139. Monrad ES, McKay RG, Baim DS, Colucci WS, Fifer MA, Heller GV, Royal HD, Grossman W. Circulation 1984;70:1030-1037.
140. Remme WJ, Wiesfeld AC, Look MP, Kruyssen HA. J Cardiovasc Pharmacol 1989;14(Suppl 2):S41-S44.
141. Warren SE, Kihara Y, Pesaturo J, Gwathmey JK, Phillips P, Morgan JP. J Mol Cell Cardiol 1989;21:1037-1045.
142. Cingolani HE, Wiedmann RT, Lynch JJ, Wenger HC, Scott AL, Siegl PK, Stein RB. J Mol Cell Cardiol 1990;22:1025-1034.

143. Kihara Y, Gwathmey JK, Grossman W, Morgan JP. Br J Pharmacol 1989;96:927-939.
144. Ferroni C, Hano O, Ventura C, Lakatta EG, Klockow M, Spurgeon H, Capogrossi MC. J Mol Cell Cardiol 1991;23:325-331.
145. Podzuweit T. In: The Regulation of Heart Function. Basic Concepts and Clinical Applications (ed.) H Rupp. Thieme Medical Publ, New York, 1986; pp. 397-406.

Conduit Coronary Artery Structure and Function

M. Gerová

Institute of Normal and Pathological Physiology, Slovak Academy of Sciences, 813 71 Bratislava, CZECHOSLOVAKIA

The first reference of the heart was found in old Surgical Papyrus from about 3000 years BC, solved by E. Smith. Coronary arteries are not mentioned there, although coronary disease was identified already in ancient Egyptian mummified hearts by Sir Ruffer in 1911 (1). Our thanks for the first drawings of "Coronales arterie" are due to Leonardo da Vinci, and to Vessalius, a great enthusiast for medicine and his illustrator, Kalkar (1).

From the aspect of PHYLOGENESIS in emerging the distinct coronary arteries, the compact layer of myocardium was shown to play a crucial role. They are to find first in fishes, some bigger amphibians and reptiles in the part of compact layer (2,3). In birds and mammals the only compact layer of the heart is nourished exclusively with distinct coronary tree (4). From aspect of ONTOGENESIS, Bogers et al (5) studying the emerging of coronary system in birds and mammals, revealed coronary arteries in human fetuses in the relatively earliest ontogenetical period in 7-8 weeks fetuses.

The development of the individual components of the coronary wall during aging is not uniform. The thickness of the media increases from the first month until adulthood from 45 u to about 200 u, that means about 4-5 times, then after this value is levelling. On the other hand, intima increases from 6 u in the first month to 300 u at adult age, that means 50 times. Moreover, the increase in thickness continues further, up to the value of 450 u - at seventies.

The parameter crucial for closing the artery in adult humans, the wall/D ratio, represents 1:7,4 (6,7). Considering that intima increases further from a value 300 u in adults to 450 u at the age of 70, and the coronary diameter does not change (6), then the geometry becomes really ominous.

To make the ontogenesis conclusive, the development of the autonomic nervous system in coronary wall is of importance. No nerve terminals either in the coronary artery or in the intracardiac small arteries in the canine fetuses, were found 1 week before birth (8). In spite, cardiomotor innervation though poor, was detected as early as 1 week before the birth. The density of nerve terminals in the myocardium was 10 times lower than in adult animals. The nerve terminals growing into the advetitia of the coronary wall were found postnatally only at the end of the first and in the second week. Thus the delay of 2-3 weeks in occurrence of nerve terminals in coronaries and myocardium has been detected.

Dynamics of Geometry of Coronary Artery Dependent on the Left and Right Ventricle Volume

The data on the geometry of coronary artery as they were outlined above were static. About 200 years after Harvey had published "De motu cordis", Purkyne lectured in Breslau on suction and pumping function of the heart: "Die Muskelsubstanz der Ventrikel nach allen Dimemsionen zusammenziehen m üsse und das Herz sich..... sowohl verkürzt, als auch verengt" (see 9). Purkyne's observation had an inevitable consequence for the biomechanics of the coronary tree, including the conduit portion.

Being in principle elastic tube and firmly tethered to the myocardium, the conduit portion of coronary tree has to be subjected to any change in the geometry of the heart so that deformation in length and diameter are to be expected.

Increasing the LV volume stepwise, from zero to a value corresponding to the normal diastolic filling, and further to a value of 100-150% higher, the segment of length of coronary artery increases about 12%, and diameter (D) decreases accordingly, i.e. about 10% (Fig. 1).

Surprisingly, was the finding that by increasing the RV-volume in

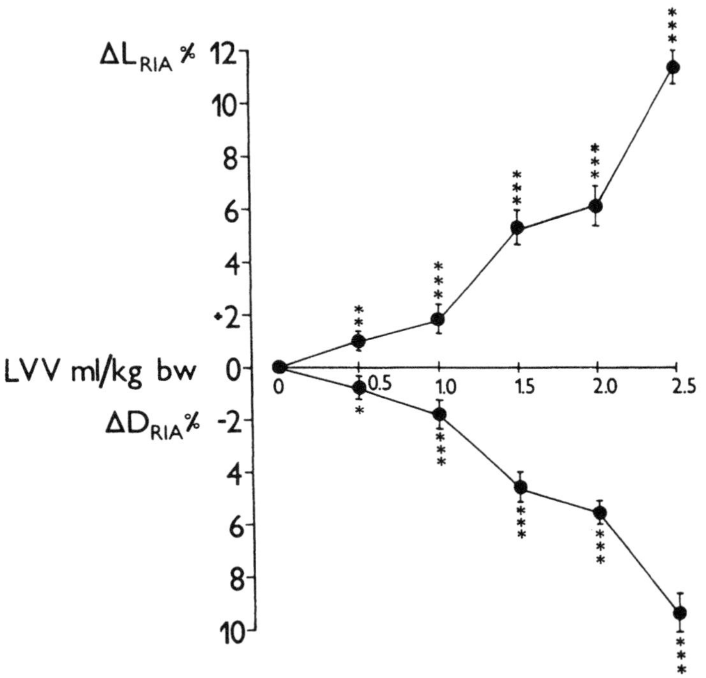

Fig. 1. Alterations in geometry of ramus interventricularis anterior (RIA) to increase in LV Volume. Abscissa: Increase in LV volume in ml normalized pro kg b.w. Ordinate: upper part: segment length increment of RIA in %, lower part: diameter decrement of RIA in %.

the same range, the ramus interventricularis anterior also alters its geometry, i.e. segment length increases and diameter decreases in spite that the vessel is running exclusively over the left ventricle. The length and diameter changes in both cases of LV- and RV- filling are quantitatively very close. The alterations in geometry along the coronary artery are not, however, uniform at the same ventricle filling: they are largest in proximal third and decrease toward apex again in both cases of LV and RV filling.

Heterogeneity in deformation was found also in the main branches of the left coronary artery: the alterations in geometry of RIA are significantly larger than of ramus circumflexus (RC). The smaller deformation of RC is a consequence of mainly two factors: (i) RC is more freely tethered; it is running loosely in the fat tissue in the atrio-ventricular groove, (ii) RC is running above almost non deformable anullus fibrosus.

In conclusion: (i) The passive deformation of RIA in length and diameter due to about 100-150% increase in L or R ventricle volume, represents an increase in resistance to flow of this particular portion from normal 10% to 23%. (ii) The deformation being transferred to individual cell components of coronary wall very probably may have consequences for the cell metabolism.

Sympathetic Control

The continuous dynamics of deformation was very probably one of the causes why the nervous control of coronary artery has not been investigated. In spite, the sympathetic and parasympathetic control of this vessel is of fundamental importance.

First, what is the morphological background of the adrenergic control for this particular segment? In dogs under aseptic conditions, the following sympathetic ganglia were extirpated: first cervical, vertebrate, stellate and three thoracic ganglia. The animals survived and 14 days after, while the postganglionic fibers degenerated, the coronary innervation was estimated. With the exception of the first cervical ganglion which does not contribute, the vertebrate, stellate and three thoracic ganglia supply the conduit portion of the coronary tree with postganglionic fibers.

Using the Falck technique, the adrenergic fibers were found spread on the surface of the adventitia and terminal fibers penetrate through it, however, they are rather far from smooth muscle cells (10).

Kristek using electron microscopy, found that there is a zone 0,5 u wide, outwards from the tunica media, where no nerve terminals are present. 50% of all terminals are spread over an area of 4,5 u, e.g. mediator released has to cover a distance of 0,5 u up to 5 u to reach the first layer of SMC. The other half of terminals is spread up to 15 u from the media and its relevancy for smooth muscle control is questionable (Fig. 2).

The varicosities contain small dense cored vesicles, empty vesicles and large dense cored ventricles, suggesting adrenergic,

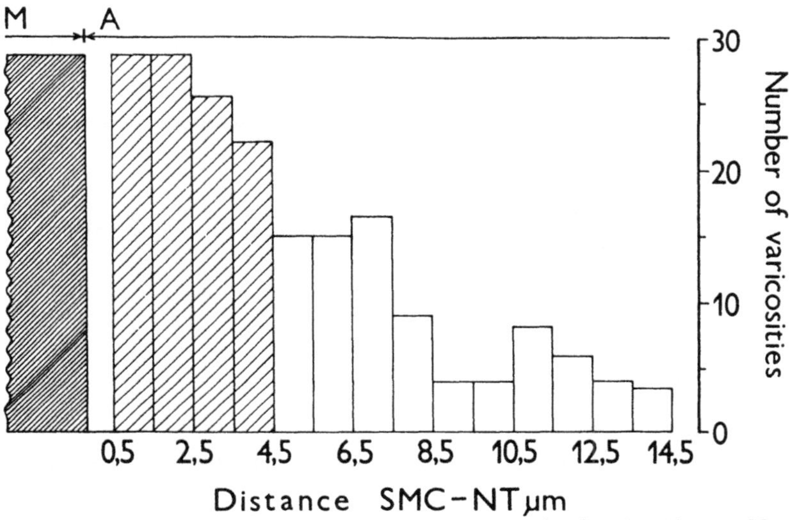

Fig. 2. Distribution of autonomic nerve terminals in the wall of RIA. From Kristek et al. (11).

cholinergic and peptidergic content. According to Franco-Cereceda (12), NPY, substance P and CGRP were found in nerve terminals of coronary arteries.

The nerve fibers running in bundles on the surface of the adventitia are easily vulnerable at coronary surgery and particularly by instrumenting the artery (13). 14 days after instrumenting the Ramus interventricularis anterior, the nerve terminals in coronary wall were distally compromised. The density of innervation in myocardium decreased by 50%. Similar injury to adrenergic innervation was found after instrumentation of RC. The density of adrenergic innervation of the myocardium decreased by 27%. The results have, of course, implications both for the practical cardiology, as well as for the experiments with chronic implanted transducers on coronary artery.

The Range of Sympathetic Control of RIA

To devoid the effect of changes in geometry of coronary arteries during sympathetic stimulation, the following experimental model was set up: In principle, the heart of the animal was arrested by

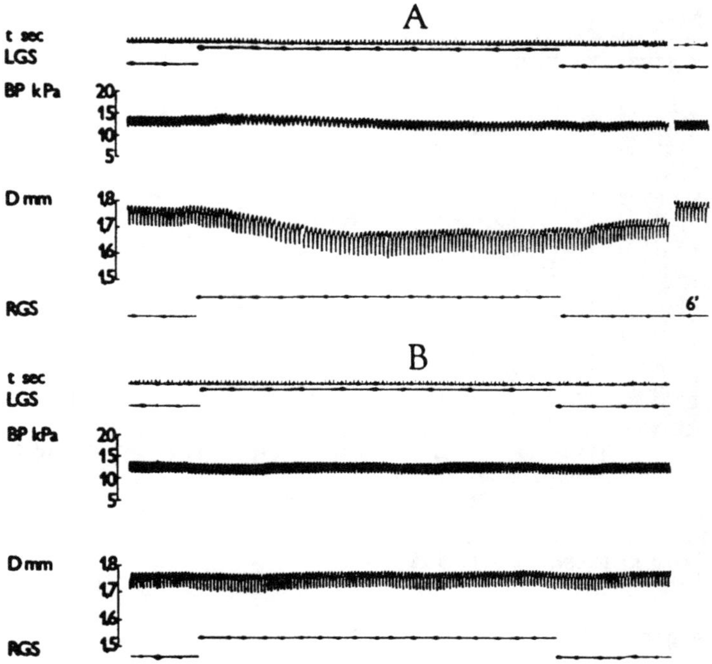

Fig. 3. The effect of left and right stellate ganglion fibers stimulation (LGS, RGS) on diameter of RIA (D mm). A before phentolamine, and B after phentolamine (2 mg/kg b.w.) was applied. From Gerová et al (14).

blocking the microcirculation with injection of mercury, and a cardiopulmonary bypass was installed, e.g. the systemic circulation, inclusively the intact innervated RIA was supplied by a roller pump, and also Bentley Temptrol oxygenator was used. Stimulating stellate ganglion bilaterally a moderate increase in BP distended lightly the coronary artery. If BP stabilizer was included in the perfusion system, the diameter of the artery decreased. This phenomenon already indicates a weak contraction of coronary SM.

The coronary constriction elicited by sympathetic stimulation represents an average 4% and- plus stimulating the three thoracic ganglia, it represents 5% of resting diameter only (14). Actually, the coronary constriction in comparison to other portions of vascular tree is one of the smallest.

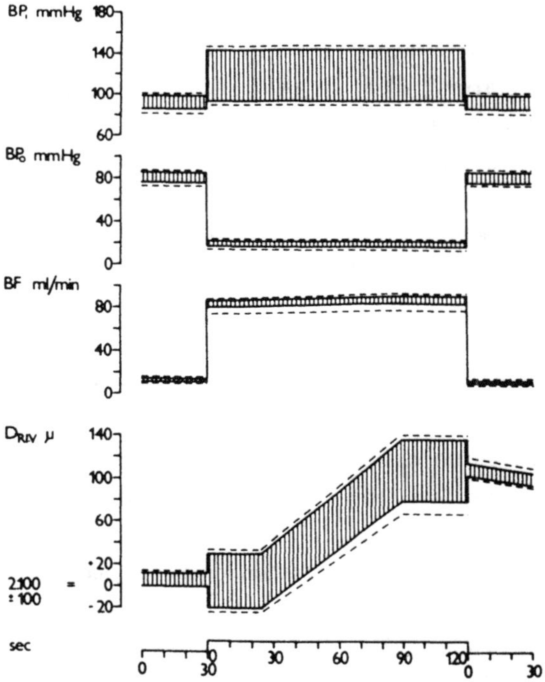

Fig. 4. The effect of increase in blood flow on RIA diameter. From top to bottom: inflow pressure (BPi), outflow pressure (BPo), blood flow (BF) and diameter (D_{RIV}). From Gerova et al. (15).

The transmitter released at the nerve terminals operates via alpha receptors because phentolamine abolished the constriction. Noteworthy is that stimulation while alpha-blockade is operating, did not elicit dilation either, e.g. beta-receptors are not involved in the response to sympathetic stimulation (Fig. 3).

To get a conclusive picture on the diameter of the coronary artery during the increase of sympathetic activity in a beating heart, at least two points have to be considered: (i) The dynamics of geometry of the heart implying the geometry of coronary artery, e.g. tachycardia, shortening of diastole, decrease in diastolic filling and/or ventricle volume with consequent decrease in segment length of coronary artery implying a passive increase in diameter. (ii) Sympathetic stimulation due to positive inochrono and dromotropic effect, increases the effective work of the heart, by this fact microcirculatory area dilates. The increase of inflow to the microcirculatory area is realized via conduit coronary artery. Thus, the question arose whether the vessel dilation due to high BF recognized by the German pathologist Thoma at the end of the last century, and repeatedly described in femoral artery by Schretzenmayer, Fleisch, Hilton and recently by a pleiade of others, is a peculiarity of femoral artery, or this mechanism is operating in coronary artery too (15).

The experiments yield a positive proof: an increase in BF through coronary artery of the dog heart, after a period of latency for about 14 sec, induced a dilation of coronary artery (Fig. 4). Shortly after, flow dilation was proven as E-dependent by Smiesko et al (16) in femoral artery, and Holtz et al (17) and Vatner (18) in coronary artery itself.

The place where the information on hemodynamics in the vessel is transferred to the smooth muscle cells and triggers the response, are the fenestrae in LEI (19). Endothelial cells or smooth muscle cells emit into fenestrae protrusions of various shape and to a various distance. Even very close junctions were found. The myoendothelial relations found admit several mechanisms of transfer of information: (i) diffusion of autocoids from EC to SMC, (ii) a direct electro-mechanical coupling between E and SMC during hemodynamical changes in the coronary artery.

Summarizing: the moderate constriction of coronary artery during sympathetic stimulation is even counter-balanced by: (i) geometrical changes of the heart, (ii) by high blood flow induced dilation which is E-dependent.

Parasympathetic Control

In spite that there are several controversial data dealing with the parasympathetic control of the coronary BF, the particular conduit portion has been hardly, if at all, studied. An indirect proof of cholinergic nerve terminals in coronary wall yield the presence of acetylcholinesterase at the adventitia-media border, detected by the Karnowsky method (20), or small clear vesicles in

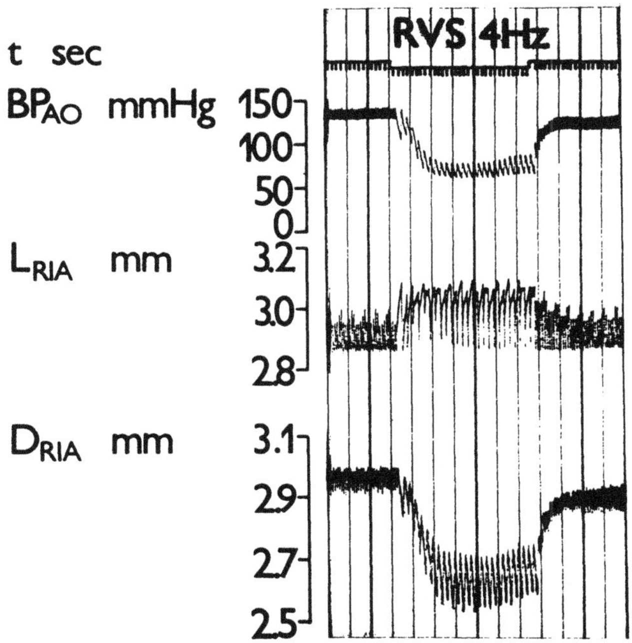

Fig. 5. Stimulation of the peripheral stump of the cut right vagus nerve (RVS) on blood pressure (BP_{AO}), segment length (L_{RIA}) and diameter (D_{RIA}) of RIA.

varicosities of nerve terminals (11). During stimulation of peripheral stump of the cut vagus nerve a deep decrease of diameter of RIA occurred (Fig. 5). Two factors contribute to the coronary diameter decrease: (i) obviously a BP decrease; (ii) the second factor is as follows: by the vagal bradycardia increases the duration of diastole, increases the enddiastolic filling and the ventricle volume, thereby the segment L of the coronary artery increases and simultaneously the diameter decreases passively. (iii) Whether there is an interfering of the third factor, a change in smooth muscle tone is not clear from these experiments.

Coronary Artery in Cardiac Hypertrophy

Dealing with the dynamics of geometry of coronary artery in

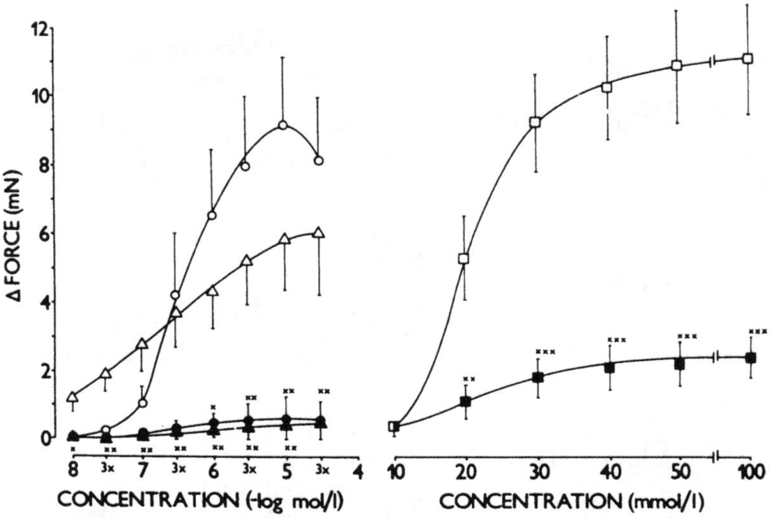

Fig. 6. Dose-response curves (isometric contractions) of coronary artery rings to serotonin O, acetylcholine △ and KCl □ , in control rabbits (open symbols) and rabbits with stabilized (4 months lasting) volume overload cardiac hypertrophy (full symbols).

relation to the left and right ventricle volume, two consequences were outlined: One was the change in resistance to BF, relevant for oxygen demand of the myocardium. The other implied the deformation of individual cells in the coronary wall. It is very probable that the continuous increase in rate of deformation of SMC in coronary wall might be a signal for metabolic alterations in coronary wall.

Question arises: How does an overload of the heart leading to hypertrophy effect the coronary artery? A volume overloading of the heart was studied and namely two aspects: structural alterations of the coronary wall and vasomotor efficiency of coronary smooth muscle.

The study was performed on a group of rabbits with experimental aortic insufficiency. After 4 months lasting volume overload inducing a stabilized cardiac hypertrophy, the heart/body weight ratio increased from 1,9 to 2,7. The thickness of coronary wall increased by 40%. The relative volume of SMC in coronary wall decreased, the extracellular space in opposite increased. Thus, the ratio SMC/ECS decreased remarkably (3,62 to 1,99). The structural

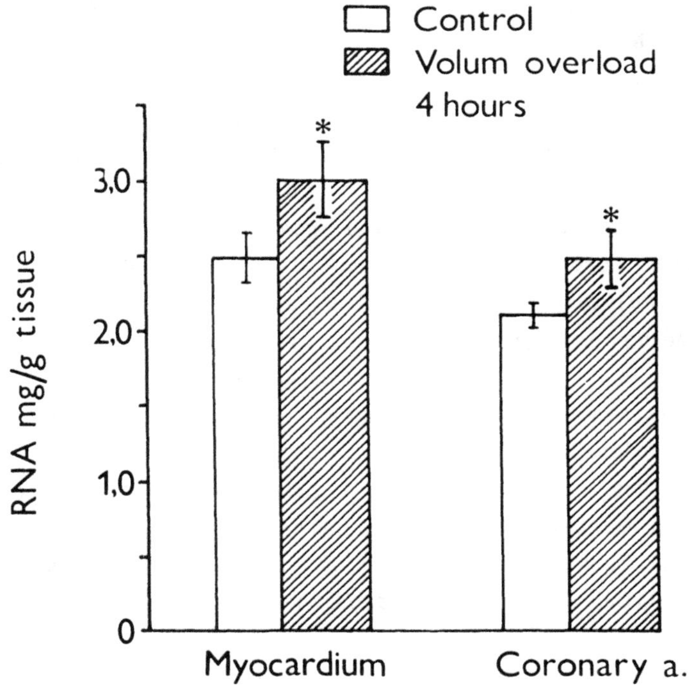

Fig. 7. RNA concentration in coronary artery and myocardium after 4 hours lasting volume overload: increase in cardiac output by 25%.

alterations found, predicted the alterations in vasomotor efficiency. Isometric contractions to serotonin, acetylcholin and potassium chloride of coronary rings from the heart with 4 months lasting volume overload, e.g. with stabilized cardiac hypertrophy, are remarkably depressed in comparison with controls (20).

The responses of coronary artery after 4 months lasting cardiac volume overload were depressed, irrelevant of mechanism triggering the constriction. This fact indicated that the key to explain the low contractile response could hardly be sought for in receptor apparatus. Considering the alterations in structure of coronary wall, it seems more likely that the SMC probably changes the phenotype, being engaged more in producing the extracellular substance in the coronary wall, and the contractile apparatus has been compromised.

Early Metabolic Processes in Coronary Wall After Pressure and Volume Overload of the Heart

There is a wealth of data maintaining that the process of cardiac hypertrophy proceeds in phases, characterized by various types of metabolism which have been explored as intensively as the expression of certain types of oncogenes. The fluctuations of metabolism in coronary wall should be expected too, however, the individual phases of response of coronary artery to the load during cardiac hypertrophy have been completely neglected. Thus, addressing the early periods of overloading two questions were raised: Does the volume or pressure overload of the heart lasting 4 hours affect the primary metabolic processes in coronary wall, namely the level of nucleic acids, and proline?

In anesthetized dogs, a shunt between left carotid artery and left auricle was performed so that cardiac output increased by 25% (22). Systolic pressure decreased significantly by 10% so that pulse pressure amplitude in aorta increased when shunt was operating by 9% significantly.

The concentration of RNA in myocardium increased significantly from 2,4 to 2,9 mg/g. What is, however, more important in connection with the question studied: after 4 hours lasting volume overload of the heart the concentration of RNA in both branches of the left coronary artery increased significantly from 2,2 to 2,5 mg/g tissue, it is about 18%. No change was found in DNA concentration and proline in coronaries, and in myocardium.

In a second series, pressure overload of the heart and coronaries was brought about by constriction of abdominal aorta above renal arteries, lasting 4 hours. This maneuver increased diastolic pressure from 120 to 165 mm Hg and systolic pressure from 144 to 211 mm Hg. The analysis of nucleic acids and proline was carried out separately in RIA and RC in these experiments, with the idea whether the differences in rate of deformation of the two branches as they were documented (9), would be reflected by the metabolism of nucleic acids.

The results obtained were interesting in several points: (i) 4 hours lasting pressure overload increased the concentration of RNA in myocardium by 13%. (ii) Concentration of RNA in RIA increased from 2,3 to 2,6 mg/g tissue that means by 12%. However, and this finding is necessary to stress, concentration of RNA in RC does not change significantly. (iii) Concentration of DNA and proline do not change either in myocardium or in coronaries.

The results of the last two series of experiments indicate that in early phases of cardiac overloading with volume or pressure, not only myocardium but also the coronary artery exhibit an increase in concentration of RNA. The RNA concentration was shown to be higher in that branch of the left coronary artery in which the rate of deformation with filling of the ventricles is higher. It seems that there would be a direct relation between the deformation of coronary wall and the turnover of RNA. Thus the idea seems to be plausible that even the short term waving of hemodynamics in coronaries implies a waving in turnover of RNA, and perhaps of other metabolic systems too.

References

1. Fishman AP, Richards DW. (eds.) In: Circulation of the Blood. Men and Ideas, Oxford University Press, New York, 1964; pp. 1-264.
2. Juhász-Nagy A, Szentivanyi M, Vámosi B. Acta Physiol Hung 1965;23:33-48.
3. Ostádal B, Rychter Z, Poupa O. Physiol Bohemoslov 1970;19:1-10.
4. Benninghoff A. In: Handbuch der Mikroskopischen Anatomie des Menschen (ed.) W von Mollendorff. Springer Verlag, Berlin 1930; pp. 217-223.
5. Bogers AJJC, Gittenberger AC, de Groot-Poelmann RE, Péault BM, Huysmans HA. Anat and Embryol 1989;180:437-441.
6. Rabe D. Basic Res Cardiol 1973;68:356.
7. Purinja B, Kasjanov BA. Biomechanics of large conduit arteries in man (in Russian) Zinatne, Riga, 1984; pp. 75-165.
8. Dolezel S, Gerová M, Hartmanová B, Vasku J. Acta Anat 1990;139:191-200.
9. Gerová M, Bárta E, Stolárik M, Gero J. Basic Res Cardiol 1989;84:583-590.

10. Dolezel S, Gerová M, Gero J, Sládek T, Vasku J. Acta Anat 1978;100:306-316.
11. Kristek F, Gerová M. Acta Anat 1987;129:149-159.
12. Franco-Cereceda P, Lundberg JM, Dahlöf C. Acta Physiol Scand 1985;124:361-369.
13. Dolezel S, Gerová M, Hartmanová B, Dostál S, Janecková H, Vasku J. Amer J Physiol 1984;246:H459-H465.
14. Gerová M, Bárta E, Gero J. Circ Res 1979;44:459-467.
15. Gerová M, Gero J, Bárta E, Dolezel S, Smiesko V, Levicky V. Bas Res Cardiol 1981;76:507-507.
16. Smiesko V, Kozík J, Dolezel S. Blood Vessels 1985;22:247-251.
17. Holtz J, Forstermann U, Pohl U, Giesler M Bassenge E. Cardiovasc Pharmacol 1984;6:1161-9.
18. Vatner S. Amer J Cardiol 1985;56:16E-22E.
19. Kristek F, Gerová M. Blood Vessels 1991; in press.
20. Gerová M, Dolezel S, Gero J, Bárta E. Physiol Bohemoslov 1979;28:299-307.
21. Holécyová A, Gerová M, Fízel A, Fízelová A. Physiol Bohemoslov 1987;36:105-111.
22. Pechánova O, Gerová M. Physiol Bohemoslov 1990;39:557.

Heart Failure: A Disease of Adaptation

B. Swynghedauw

Unité 127 INSERM, Hopital Lariboisiere, 41 Bd de la Chapelle, 75010 Paris, FRANCE

Modern biology has not transformed the landscape of cardiac insufficiency, CI, but to some degree in this rather complex domain, it has permitted us to actually distinguish between what is and what is not a physiological process of adaptation of a diseased organ. This distinction is not one of rhetorical importance. It facilitates the dialogue between clinicians and fundamentalists. Above all it permits us to look at therapy from a different perspective and hopefully at the conception of new drugs. From this progress has emerged evidence that the structure of inotropic targets, myocardial contractile cells, are already profoundly modified at the moment when the first peripherial clinical signs that define CI arise. In other terms, inotropic agents, more or less those drugs having a myocardial activity, are advisably the keys to use in conceptualizing function because the lock in which they must play is modified [1].

Perhaps it is no longer necessary to summarize (Table 1) a number of the confusions, the most habitual of which transforms the exchanges in the course of such a scientific meeting to muffled dialogues. That which follows will attempt to take the part between the inappropriateness of terms and really scientific distinctions.

The Principal Parameters to the Origin of C.I.

Certainly on this point, there is one major source of incomprehension between the different partners concerned with this ailment. We can, in an effort for clarification, propose to distinguish four groups of parameters (Table 2), beginning with those that are, when they exist, the most clinically pronounced, and terminating with those that are always biologically speaking present although often of minor clinical expression, that is to say the physiological processes by which the heart as an organ adapts to the modifications of work, that is its environment.

Etiologic factors, and in particular coronary insufficiency can play a major role. In 1989, schematically CI is a disease of the sixties that complicates an arterial hypertension associated with a coronary insufficiency. For this last etiology the loss of substance plays a pronounced role that can push to the background all the other factors. Fibrosis or other resulting etiologies play an important role in the determination of CI in man that experimenters who make the models more simple have a tendency to forget.

The deleterious myocardial effects of neuroendocrine control. Secretions of catecholamines, vasopressin, angiotensin II or atrial natriuretic factor, like the baroreflexes, aim to maintain an organ perfusion when the hemodynamic conditions begin to be altered. They do not have beneficial effects. Besides their arrhythromogenic effects, catecholamines are the origin of multiple necrotic foci, and angiotensin II is probably one of the determinants of vascular hypertrophy. The advent of specific therapies demonstrated, which are detailed on several occasions in this book, that the correction of this initially adaptational process finally had a beneficial effect.

The adaptational process is limited and imperfect. In response to mechanical overload the expression of the cardiac genome is modified at once quantitatively, as with hypertrophy, and qualitatively.

The limits of adaptation are anatomical and biological, at the level of the myocardium and the peripery. Hypertrophy is adaptational because the number of contractile elements multiply and because it corrects the wall stress, but is limited by anatomical considerations and because in the adult, the myocyte can no longer divide.

Modern biology has demonstrated qualitative changes in genetic expression in animals and in man. They account to some degree for the qualitative changes of the contractile properties of the hypertrophied tissues and finally permit the hypertrophied cardiac fibers to develop a normal active tension at the expense of speed. The process is moreover exactly the same as that used by skeletal muscle, which obviously has an advantage over myocardial muscle since it is always submitted to an intermittent mechanical overload. One of the better documented examples in the rat ventricle, as in the human atria, is that of the myosin isoenzymes. Under the effects of mechanical overload, the majority of myosin is composed of a slow myosin isoform, V3, and it is certain that this particular adaptational process has reached its limit.

One of the stronger arguments that permits one to affirm that the physiological process of adaptation has limits beyond which to release itself from CI, besides simply good sense, is the fact that CI can arise in simple experimental models without any ultrastructural myocyte change, without necrosis, or without irreversible biochemical signs of ischemia.

Imperfections of the adaptational process are foreseeable and evident. Foreseeable because in nature there are no perfect adaptational processes, no more in the course of evolution than within a given species. Polycythemia and arterial hypertension that structurally adapt the organism to altitude have some disadvantages just as those seen in the immunological depression that adapts the mother to the fetal graft. "Adaptation corresponds more to a series of modifications whence the sum is favorable to that as a veritable response to a physiologic problem". The same can be seen in the course of evolution, the evolutionary choice permits a species to reproduce and develop, it is never a perfect response to the exact problem raised by the environment.

The imperfections in our system are numerous. (i) The slowing of Vmax, an adaptational process, indispensable at the level of the fibers, has some obvious deleterious consequences at the level of cardiac flow. (ii) Subendocardial angiogenesis is not activated in the same manner as the whole of protein synthesis. (iii) The increase in muscular mass, as in the quantitative and qualitative changes of collagen, plays a determining role in the genesis of compliance abnormalities and in diastolic dysfunction. (iv) Although this may not be perfectly clear, action potential and the entering calcium current modifications seem to some degree, to account for the spontaneous arrhythromogenicity of myocyte hypertrophy. All this has been detailed previously.

The Adaptational Process

From a structural point of view, the idea that the skeletal muscle adapts to become more resistant to fatigue, is now admitted as much by molecular biologists as by sport medicine specialists. It is always surprising to see the resistances that are met when one proposes to apply this same concept in the same manner to the myocardium.

A number of important experimental data however permits one to affirm that in response to a permanent or intermittent mechanical overload, muscle, more or less striated muscle but also very probably smooth muscle, changes in structure, the result of which permits the fiber to continue contracting albeit more slowly to ensure its normal function, that is to say the muscle develops an active tension under the best possible thermodynamic conditions.

It is not necessary to develop a complex mechanical approach to understand because cardiac papillary muscle induced to lift an afterload greater than that to which it is accustomed, instantaneously contracts at a speed slower than normal. The immediate consequence, demonstrated experimentally [1], will be a fall in the economy of the system, with the muscle using more ATP (or oxygen) per gram of developed tension than normal. This can be analogously expressed without hazard, by mechanics. An automobile has an optimal speed, a speed for which the yield, economy, or if one prefers, the number of liters of gasoline burned per km is maximal. On either side of this optima the yield diminishes.

Changes in the biological structure of the myocardium are going to permit this to continue to contract in an economic way using another speed of shortening/post charge curve. Obliged to role the vehicle more slowly, if it wants to continue to role economically with an increased load, it will gradually change its motor.

As one measures the accumulation, that which piles up, and the given supplies by physiology and by cellular biology it appears more and more clear that what permits the survival of the mechanical organ, is the slowing of the maximal velocity of shortening.

In cellular physiology the duration of the action potential, as the length of intracellular variations (transitions) of calcium concentrations are slower. Relaxation is slower and above all the heat production per gram of developed tension is at this time decreased and prolonged.

Cellular biology, one has seen it equally higher, begins to explain a large part of these changes, among others those of relaxation. The biological reason for these physiological changes are nevertheless species specific, all as the normal metabolism of calcium. In rat, the slowing of Vmax is explained to some degree by a change in the structure of the calcium "receptors" i.e., the contractile proteins. In man, guinea pig, cat, dog and pig, the situation is quite different since the ventricular contractile proteins do not change; therefore, it is necessary to admit that the determining changes of Vmax are localized at the membrane protein level responsible for intracellular calcium movements.

The signals that inform the cardiac genome of mechanical modifications are in the domains of research that have recently experienced explosive developments. Stretching out ("stretch"), the only factor known for certain, isometric tension and thermodynamic modifications are 3 possible candidates. The transmission of these signals to the genome occurs by an unknown process, the best candidate for the moment being the cytoskeleton. Molecular biology has by contrast learned that the reaction of the genome is double and understands the real transitory signals without adaptational significance and those that could prepare the gene for permanent adaptational changes.

Treating Cardiac Insufficiency Accordingly

The success and the failures of some inotropes, vasodilators and

Table 1. Sources of confusion in understanding cardiac insufficiency, C.I.

- The distinction between "physiologic" and "pathologic" hypertrophy
- To assimilate chronic and acute C.I., mechanical overload and cardiomyopathy
- To describe as "abnormal" the biochemical changes or the modifications of genetic expression
- To consider the inotropic treatment alone, or the treatment by vasodilators alone, as dogma
- To look for a specific biologic marker of cardiac failure

Table 2. Determining factors of systolic ejection in cardiac insufficiency of coronary origin.

i) At the level of healthy tissue: factors directly relevant to the physiologic adaptive process:
- concentric compensatory hypertrophy (secondary to stretch) and eccentric compensatory hypertrophy (Starling's law)
- shortening of Vmax

ii) Habitual imperfections of the adaptive process:
- at the level of normal tissue, insufficient subendocardial angiogenesis
- filling abnormalities: shortening of the speed of relaxation, less compliance of hypertrophied tissue

iii) Associated factors:
- etiologic: loss of substance; rigidity of scar tissue; depending on animal species or more probably on the size of the infarct, decrease or increase in chamber rigidity; asymmetry at the time of the cells and the organ, source of arrhythrogenicity and mechanical dysfunction
- neuroendocrine: angiotensin II (can be an activator of the gene for collagen), catecholamines

Table 3.

i) Natural history of the adaptive process:
- change the genomic expression (thyroxine analogues)
- change the load (vasodilators, diet, diuretics)

ii) Correct the imperfections of the adaptive process:
- simulate angiogenesis
- is it necessary to accelerate Vmax (?? distinguish between inotropes and inotropic effects)
- facilitate filling (bradycardic agents, relaxing effects of beta adrenergic agonists, stimulate the expression of the SR genes ?)

iii) Take into account associated factors:
- correct the loss of substance (graft, cardiomyoplasty)
- put an end to excess neuroendocrine reactions: IEC, anti-renin, saralasin, beta adrenergic antagonist, antivasopressin

the diuretics have been previously detailed. To modify or improve the natural adaptational process is obviously the primary objective. Vasodilators and diuretics do not themselves modify the adaptational process, but they lower the adaptational threshold and bring it back under the threshold that has been passed beyond which the myocyte has shown that it was less adaptable. Their effect on survival has more or less been demonstrated. It is nevertheless unthinkable to conceive of a therapy modifying the genomic expression that is by definition perfect. Thyroxine or its analogs, some of which are specific for certain tissues, undoubtedly modify the expression of the cardiac genome. Completed clinical trials are rare but among them one has had beneficial effects, perhaps because the hormone stimulates the synthesis of adrenergic receptors.

Another aspect of the problem, is the correction of the systems obvious imperfections. Specific factors of angiogenic growth are not now in the domain of the unreal, well to the contrary. More realistically, is to rethink the inotropes. It is in fact difficult to accept biologically that accelerating Vmax will be beneficial, and it is probably more reasonable to think that the beneficial effects of the inotropic drugs are not due to their inotropic effects but because they facilitate the filling of the heart by slowing the frequency, like digitalin, or by accelerating relaxation, like beta adrenergic agonists. It is not reasonable in all cases to bury this therapeutic family in the name of clinical trials, contested, or in the name of modern biology, to the profit of vasodilators, less inoffensive than we say it too often, the nephrologic medications can witness some.

To hold in account the associated factors is the last horizon. Cardiomyoplasty, using a skeletal muscle previously transformed, is one of the best examples of physiology applied in this domain. It is equally necessary to underline that those tests, just now fruitful, that stop the excess neuroendocrine reaction have not yet all been attempted.

Actual biology leads, in a paradoxical way, to a large modesty. It is certain that in some manner it permits a justification of such or such a gospel concerning the therapy of cardiac insufficiency. If we hold it as an account of a certain number of economic parameters that are far from innocent in this domain, it is actually not reasonable to attribute to the vasodilators a clearly superior efficacy to that of the inotropes particularly since in the final analysis it seems necessary to establish a distinction between inotropes and inotropic effects.

Reference

1. Swynghedauw B. Cardiac hypertrophy and failure. INSERM J 1990; Libbey, Paris, London.

Involvement of Sarcolemmal Na^+-K^+ ATPase in the Pathogenesis of Heart Disease

N.S. Dhalla, Q. Shao and I.M.C. Dixon

Division of Cardiovascular Sciences, St. Boniface General Hospital Research Centre, University of Manitoba, Winnipeg, CANADA

Introduction

It is now well established that Na^+-K^+ ATPase is primarily localized in the sarcolemmal membrane in the myocardium (1,2). This enzyme is intimately involved in the transport of Na^+ and K^+ and is thus considered to maintain the electrochemical gradient across the cell membrane which allows depolarization to occur. An inhibition of Na^+-K^+ ATPase activity has been shown to increase the intracellular concentration of Na^+, which then increases the cytoplasmic level of Ca^{2+} through Na^+-Ca^{2+} exchange system in the plasma membrane and thus has been proposed to serve as a mechanism of the positive inotropic action of cardiac glycosides (digitalis). Although Na^+-K^+ ATPase is known to possess high and low affinity sites for cardiac glycosides, the exact significance of these sites is not clear at present (3). Since the intracellular concentration of Na^+ has been shown to regulate myocardial contractility (4), any change in the operation of sarcolemmal Na^+-K^+ ATPase activity can be seen to alter Ca^{2+} ion movements in the cell and thus modify the heart function. Accordingly, a great deal of attention has been focussed on elucidation of the function of this enzyme in both health and disease.

Recently, it has been demonstrated that Na^+-K^+ ATPase is coupled with G proteins, which may serve as transducers for hormonal inhibition of this enzyme system in the myocardial sarcolemmal membrane (5). Up-regulation of the Na^+-K^+ ATPase in the myocardium has been shown to occur after chronic administration of digitalis (6) as well as 7-oxo-prostacyclin (7). The enzyme consists of a catalytically active alpha-subunit with a molecular wt of about 112 KD and a glycosylated beta-subunit with a molecular wt of about 35 KD (8,9). In adult myocardium, two isoforms ($alpha_1$ and $alpha_2$) of the Na^+-K^+ ATPase alpha-subunit have been shown to be expressed. Selective alterations in Na^+-K^+ ATPase isoforms have also been reported to occur in cardiac muscle in K^+-deficiency (10). Several studies have revealed that the increased activity of myocardial Na^+-K^+ ATPase due to thyroid hormones may be due to a differential change in various isoforms of the Na^+-K^+ ATPase (11-16). Since thyroid hormones are known to increase the rate of contraction, rate of relaxation and heart rate, it is possible that these effects of the hormone are elicited by changes in the composition of Na^+-K^+ ATPase isoforms. However, much work in this regard remains to be carried out prior to reaching any meaningful conclusion. Furthermore, a wide variety of changes in myocardial Na^+-K^+ ATPase have been identified in different types of heart disease (as described below) but very little is known with regard to the molecular mechanisms for these alterations.

Cardiac Insufficiency and Cardiomyopathy

In light of our existing knowledge concerning the actions of cardiac glycosides on heart function and Na^+-K^+ ATPase, inhibition of this enzyme should lead to an increase in cardiac contractility and vice versa. This view is consistent with the finding that increased Na^+-K^+ ATPase activity was seen in cardiac insufficiency in rats induced by vitamin E deficiency (17), administration of cobalt chloride (18), or 20,25-diazocholesterol (19). Rats fed a high cholesterol diet for 6 to 12 weeks also showed an increase in heart sarcolemmal Na^+-K^+ ATPase activity (20). Prolonged treatment of animals with digitoxin or potassium-deficient diet (21) resulted in an increase in Na^+-K^+ ATPase activity in the heart. In the BIO 14.6 strain of cardiomyopathic hamsters, Na^+-K^+ ATPase activity was increased in moderate and severe stages of cardiomyopathy (22-25). However, depressed Na^+-K^+ ATPase activity has been reported in diseased hearts from UM-X7.1 strain of cardiomyopathic hamsters (26-28). Furthermore, heart dysfunction in chronic diabetes in rats was also associated with decreased sarcolemmal Na^+-K^+ ATPase activity (29). These changes cannot be attributed to the cardiomyopathic process per se because no changes in myocardial Na^+-K^+ ATPase were seen upon inducing cardiomyopathy with high doses of catecholamines in rats (30). Thus, in view of the elevated and depressed activities of the sarcolemmal Na^+-K^+ ATPase in various models of cardiac insufficiency and cardiomyopathy, it is difficult to explain diminished contractile force in chronically diseased hearts on the basis of conventional interpretation of the enzyme function under acute situations. Nonetheless, it is becoming evident that changes in sarcolemmal Na^+-K^+ ATPase activity are dependent upon the stage and type of heart disease.

Myocardial Ischemia and Reperfusion

Several studies over the past 25 years have revealed that reduction in coronary flow (myocardial ischemia) results in the impairment of contractile function, myocardial cell damage, electrolyte imbalance and metabolic derangement (31,32). Furthermore, if reperfusion of the ischemic myocardium is not instituted within a certain time, additional defects (reperfusion injury) become evident (33). Such defects in the dog heart due to ischemia-reperfusion injury have been associated with depressed sarcolemmal Na^+-K^+ ATPase activity (34,35). Human cardiac tissue from patients with atherosclerotic coronary artery disease also showed decreased Na^+-K^+ ATPase activity (36). Since both oxygen-lack and substrate-lack are considered to play an important role in the ischemic heart disease, a depression in Na^+-K^+ ATPase activity was seen in isolated hearts perfused with hypoxic or substrate-free medium (24,37-40). A decrease in sarcolemmal Na^+-K^+ ATPase activity was also seen in Ca^{2+}-paradoxic heart in which intracellular Ca^{2+}-overload, like that seen in the ischemic-reperfused heart, is considered to result in myocardial

Table 1. General characteristics, hemodynamics and heart sarcolemmal ATPase activities in control and experimental rats 8 weeks after occluding coronary artery.

	Control	Experimental
Body wt (g)	505 ± 8.03	501 ± 10.02
Viable left ventricle wt (mg)	0.93 ± 0.78	1.32 ± 0.49*
Ascites (ml)	---	3.53 ± 0.12*
Lung wet/dry wt	3.82 ± 0.31	5.21 ± 0.26*
Right ventricle wt/body wt (mg/g)	6.48 ± 0.04	0.76 ± 0.03*
LVSP (mm Hg)	121 ± 7.42	112 ± 6.90
LVEDP (mm Hg)	2.13 ± 0.91	13.1 ± 1.24*
+ dP/dt (mm Hg/s)	5580 ± 250	4012 ± 108*
- dP/dt (mm Hg/s)	5248 ± 236	3260 ± 225*
Sarcolemmal yield (mg/g)	1.75 ± 0.21	1.86 ± 0.16
Na^+-K^+ ATPase (umol/mg/hr)	20.5 ± 1.42	12.0 ± 0.78*
Ca^{2+}-stimulated ATPase (umol/mg/hr)	22.4 ± 1.63	21.8 ± 1.46

Values are means ± S.E. of 8 experiments. LVSP and LVEDP are left ventricular systolic and end-diastolic pressures, respectively. + dP/dt indicates the rate of contraction wherease - dP/dt represents the rate of relaxation of the left ventricle. Only the viable left ventricular tissue was used for the isolation of sarcolemma. * $P < 0.05$ from respective control.

abnormalities. In fact perfusion of the infarcted-reperfused heart with low Ca^{2+} has been shown to prevent the depression of Na^+-K^+ ATPase activity (41). In view of the formation of oxygen free radicals and other oxidants as well as accumulation of different lipid metabolites in the ischemic-reperfused myocardium, various investigators have examined the role of these pathogenetic factors in

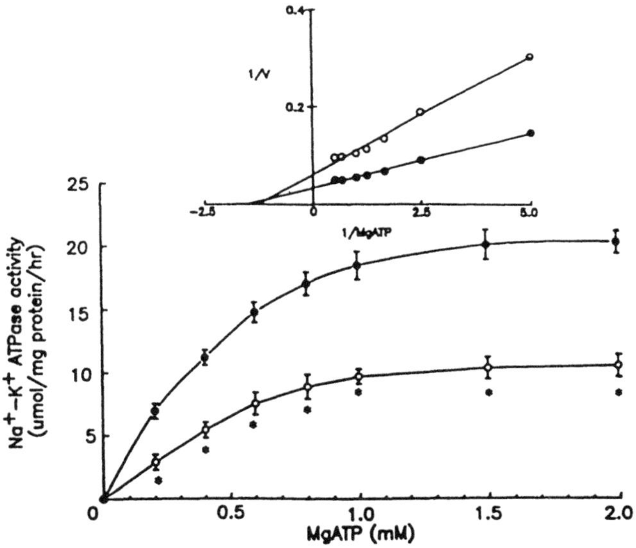

Figure 1. Saturation kinetics of Na^+-K^+ ATPase activity with MgATP in sarcolemmal preparations from control and experimental (8 weeks following induction of myocardial infarction) left ventricle. Inset: Lineweaver-Burk plot of the data. Control - ●--●; failing -O--O; * $P < 0.05$ when compared to the control values. Each value is a mean of six experiments.

inducing changes in the sarcolemmal Na^+-K^+ ATPase. Accordingly, varying degrees of depression in Na^+-K^+ ATPase activity has been observed upon exposing the heart sarcolemmal membranes to different species of oxygen free radical generating systems and oxidants (42-46). Likewise, different studies have reported a marked depression in the myocardial sarcolemmal Na^+-K^+ ATPase upon incubation with various long chain lipid metabolites (47-52). From these studies it appears that myocardial ischemia-reperfusion results in depressing the sarcolemmal Na^+-K^+ ATPase activity; however, neither the mechanisms nor the significance of this change in the ischemic heart disease are clearly understood.

Cardiac Hypertrophy and Congestive Heart Failure

Congestive heart failure is invariably preceded by cardiac hypertrophy, thus some attempts have been made to assess the status

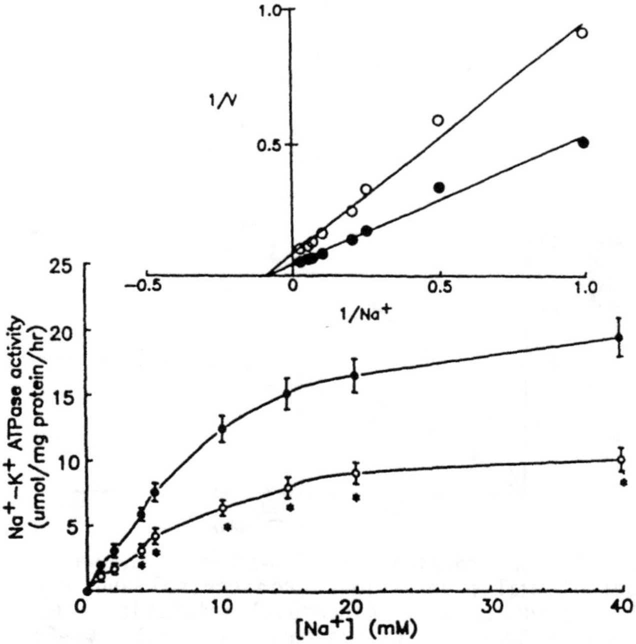

Figure 2. Response of Na^+-K^+ ATPase activity to Na^+ ion concentration in sarcolemmal preparations from control and experimental (8 weeks after induction of myocardial infarction) left ventricle. Inset: Lineweaver-Burk plot of the data. Control - ●--●; failing - O--O ; * $P < 0.05$ when compared to the control values. Each value is a mean ± SEM of six experiments.

of sarcolemmal Na^+-K^+ ATPase during the development of heart hypertrophy and congestive heart failure (53,54). Although elevated Na^+-K^+ ATPase activity was observed in hearts of dogs with failing left ventricle due to mitral valve insufficiency (55) or aortic constriction (56), depressed Na^+-K^+ ATPase activity was also demonstrated in left ventricular hypertrophy and failure in rabbits with aortic constriction (57). At present it is difficult to explain these conflicting results. Nonetheless, the observed change in Na^+-K^+ ATPase cannot be considered to be due to myocardial hypertrophy per se because no alterations in the sarcolemmal Na^+-K^+-ATPase were seen in rabbits with stable cardiac hypertrophy

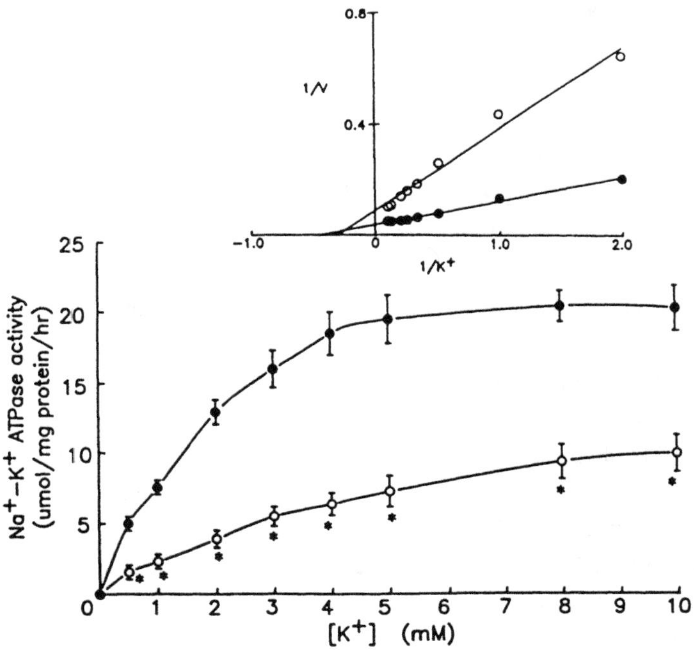

Figure 3. Response of Na^+-K^+ ATPase activity to K^+ ion concentration in sarcolemmal preparations from control and experimental (8 weeks after induction of myocardial infarction) left ventricle. Inset: Lineweaver-Burk plot of the data. Control - ●--●; failing - O--O; * $P < 0.05$ when compared to the control values. Each value is a mean ± SEM of six experiments.

(58). Myocardial hypertrophy and congestive heart failure due to bacterial endocarditis in rabbits (59) as well as due to myocardial infarction (60) were associated with a marked depression in Na^+-K^+ ATPase activity. Although the exact mechanisms for changes in sarcolemmal Na^+-K^+ ATPase in failing hearts are not clear, recent studies employing molecular biotechnology have revealed switching of Na^+-K^+ ATPase isoforms in cardiac hypertrophy and heart failure (61-63).

In our efforts to investigate changes in the characteristics of sarcolemmal Na^+-K^+ ATPase in congestive heart, we employed a rat model of congestive heart failure due to myocardial infarction

(60,64,65). Some of the hemodynamic data in rats 8 weeks following coronary occlusion are given in Table 1. On the basis of clinical signs such as formation of ascites, lung congestion (as reflected by lung wet/dry wt ratio) as well as marked changes in the hemodynamic parameters and in view of our previous work (60,64,65), the animals at 8 weeks with more than 30% infarction of the left ventricle are considered to be at a moderate stage of congestive heart failure. The sarcolemmal Na^+-K^+ ATPase in the non-ischemic viable left ventricle from experimental animals was found to be depressed (Table 1). This change was specific in nature since the activity of another ATPase, Ca^{2+}-pump ATPase, in the failing heart was not different from the control preparation. When the Na^+-K^+ ATPase activity was determined in the presence of different concentrations of MgATP, the depression in the enzyme activity was not associated with any alterations in the affinity of the enzyme for its substrate (Fig. 1). Likewise, no changes in the apparent dissociation constants were observed when the measurement were made in the presence of different concentrations of Na^+ (Fig. 2) or K^+ (Fig. 3) whereas depression in the enzyme activity was seen at each concentration of these cations. Although it is assumed that this depression in Na^+-K^+ ATPase activity is due to a switch in Na^+-K^+ ATPase isoforms, the exact mechanisms for this change remain to be investigated. Earlier studies have suggested that this change may serve as an adaptive mechanism during the congestive heart failure (60); however, the significance of depressed Na^+-K^+ ATPase in congestive heart failure needs to be established.

Summary and Conclusion

On the basis of the existing information in the literature, it is evident that sarcolemmal Na^+-K^+ activity is either increased or decreased in different experimental models of cardiac contractile failure. It is suggested that these changes in this enzyme activity are dependent upon the stage and type of heart disease. Although the molecular mechanisms for changes in myocardial Na^+-K^+ ATPase are not clear, the possibility of switching isoforms for this enzyme seems quite attractive. Nonetheless, congestive heart failure due to myocardial infarction appears to result in the depression of sarcolemmal Na^+-K^+ ATPase and this may serve as an adaptive mechanism for the maintenance of contractile activity in the failing heart.

Acknowledgements

The research reported in this paper was supported by a grant from the Heart and Stroke Foundation of Manitoba.

References

1. Schwartz A, Lindenmayer GE, Allen JC. Pharmacol Rev 1975;27:3-134.
2. Schuurmans-Stekhoven F, Bonting SL. Physiol Rev 1981;61:2-76.
3. Mogul D, Rasmussen HH, Singer DH, Ten Eick RE. Circ Res 1989;64:1063-1069.
4. Sheu SS. In: Inotropic stimulation and myocardial energetics (eds.) H Just, C Holubarsch and H Scholz. Springer-Verlag, New York 1989;35-45.
5. Danilenko MP, Turmukhambetova VC, Yesirev OV, Tkachuk VA, Panchenko MP. Am J Physiol 1991;261 Suppl:87-91.
6. Bonn R, Greeff K. Arch Int Pharmacodyn Ther 1978;233:53-64.
7. Dzurba A, Ziegelhoffer A, Breier A, Vrbjar N, Szekeres L. Cardioscience 1991;2:105-108.
8. McDonough AA, Schmitt C. Am J Physiol 1985;248:C247-C251, 1985.
9. Sweadner KJ, Farshi SK. Proc Natl Acad Sci USA 1987;84:8404-8407.
10. Azumua KK, Hensley CB, Putman DS, McDonough AA. Am J. Physiol 1991;260:C958-C964.
11. Chaudhury S, Ismail-Beigi F, Gick GG, Levenson R, Edleman IS. Mol Endocrinol 1987;162:156-159.
12. Gick SS, Melikian J, Ismail-Beigi F. J Membr Biol 1990;115:273-282.
13. Horowitz B, Hensley CB, Quintero M, Azuma KK, Putman D, McDonough AA. J Biol Chem 1990;265:14308-14314.

14. Klein LE, Bartolomei BMS, Lo CC. Am J Physiol 1984;247:H570-H575.
15. Ng YC, Yao AZ, Akera T. Am J Physiol 1989;257:H534-H539.
16. Hensley CB, Azuma KK, Tang KJ, McDonough AA. Am J Physiol 1992;262:C484-C492.
17. Fedelesova M, Sulakhe PV, Yates JC, Dhalla NS. Can J Physiol Pharmacol 1971;49:909-918.
18. Dransfeld H, Lipinski J, Borsch-Galetke E. N-S Arch Pharmacol 1971;270:335-342.
19. Fehn W, Seiler D, Kuhn E, Bartels D. Europ J Clin Invest 1975;5:327-330.
20. Moffat M, Dhalla NS. Can J Cardiol 1985;1:194-200.
21. Bluschke V, Bonn R, Greeff K. Europ J Pharmacol 1973;37:189-191.
22. Dhalla NS, Sulakhe PV, Fedelesova M, Yates JC. Adv Cardiol 1974;13:282-300.
23. Appelt AW, Welty JD, Peterson MB. J Mol Cell Cardiol 1976;8:901-907.
24. Dhalla NS., Tomlinson CW, Singh JN, Lee SL, McNamara DB, Harrow JAC, Yates JC. Rec Adv Stud Card Struc Metab 1976;9:377-394.
25. Sulakhe PV, Dhalla NS. Exp Mol Pathol 1973;18:100-111.
26. Dhalla NS, Singh JN, Bajusz E, Jasmin G. Clin Sci Mol Med 1976;51:233-242.
27. Singh JN, Dhalla NS, McNamara DB, Bajusz E, Jasmin G. Rec Adv Stud Card Struc Metab 1975;6:259-268.
28. Panagia V, Singh JN, Anand-Srivastava MB, Pierce GN, Jasmin G, Dhalla NS. Cardiovas Res 1984;18:567-572.
29. Pierce GN, Dhalla NS. Am J Physiol 1983;245:C241-C247.
30. Makino N, Dhruvarajan R, Elimban V, Beamish RE, Dhalla NS. Can J Cardiol 1985;1:225-232.
31. Dhalla NS, Das PK, Sharma GP. J Mol Cell Cardiol 1978;10:363-385.
32. Dhalla NS, Panagia V, Singal PK, Makino N, Dixon IMC, Eyolfson DA. J Mol Cell Cardiol 1988;20 (Suppl II):3-13.
33. Sharma GP, Varley KG, Kim SW, Barwinsky J, Cohen M, Dhalla NS. Am J. Cardiol 1975;36:234-243.

34. Beller GA, Conroy J, Smith TW. J Clin Invest 1976;57:341-350.
35. Schwartz A, Wood JM, Allen JC, Bornet EP, Entman ML, Goldstein MA, Sordahl LA, Suzuki M. Am J Cardiol 1973;32:46-61.
36. Lindenmayer GE, Sordahl LA, Harigaya S, Allen JC, Besch HR, Schwartz A. Am J Cardiol 1971;27:277-283.
37. Balasubramanian V, McNamara DB, Singh JN, Dhalla NS. Can J Physiol. Pharmacol 1973;51:504-510.
38. Dhalla NS, Singh JN, Fedelesova M, Balasubramanian V, mcNamara DB. Cardiovas Res 1974;8:227-237.
39. Dixon IMC, Eyolfson DA, Dhalla NS. Am J Physiol 1987;233:H1026-H1034.
40. Dhalla NS, Dixon IMC, Beamish RE. Biomed Biochem Acta 1987;46:S505-S511.
41. Samovilidon E, Levis GM, Darsinos JT, Pistevos A, Karli JN, Tsiganos CP. Biochim Biophys Acta 1991;1070:343-348.
42. Kramer JH, Mak IT, Weglicki WB. Circ Res 1984;55:120-124.
43. Kim MS, Akera T. Am J Physiol 1987;252:H252-H257.
44. Xie Z, Wang Y, Askari A, Huang W-H, Klaung JE, Askari A. J Mol Cell Cardiol 1990;22:911-920.
45. Kukreja RC, Weaver AB, Hess ML. Am J Physiol 1990;259:H1330-H1336.
46. Vinnikova AK, Kukreja R, Hess ML. J Mol Cell Cardiol 1992;24:465-470.
47. Adams RJ, Cohen DW, Gupta S, Johnson JD, Wallick ET, Wang T, Schwartz A. J Biol Chem 1979;254:12404-12410.
48. Lamers JMJ, Stinis HT, Montfoort A, Hulsmann WC. Biochim Biophys Acta 1984;774:127-137.
49. Kakar SS, Huang W-H, Askari A. J Biol Chem 1987;262:42-45.
50. Abe M, Yamazaki N, Suzuki Y, Kobayashi A, Ohta H. J Mol Cell Cardiol 1984;16:239-245.
51. Dhalla NS, Kolar F, Shah KR, Ferrari R. Cardiovasc Drug Therap 1991;5:25-30.
52. Pitts BJR, Okhuysen CH. Am J Physiol 1984;247:H840-H846.
53. Dhalla NS, Heyliger CE, Beamish RE, Innes IR: Can J Cardiol 1987;3:183-196.

54. Lamers JMJ, Stinis JT, Kort WK, Hulsmann WC. J Mol Cell Cardiol 1978;10:235-248.
55. Khatter JC, Prasad K. Cardiovasc Res 1976;10:673-641.
56. Prasad K, Khatter JC, Bharadwaj B. Cardiovasc Res 1979;13:95-104.
57. Yazaki Y, Fujii J. Jph Heart J 1972;13:73-83.
58. Heyliger CE, Dhalla NS. J Appl Cardiol 1986;1:447-467.
59. Tomlinson CW, Lee SL, Dhalla NS. Circ Res 1976;39:82-92.
60. Dixon IMC, Hata T, Dhalla NS. Am J Physiol 1991;262:C664-C671.
61. Herrera VLM, Ruiz-Opazo N. Science 1990;249:1023-1025.
62. Tsuruya Y, Ikeda U, Kawakami K, Nagano K, Kamitani T, Oguchi A, Ebata H, Shimada K, Medford RM. Clin Exp Hyper Theo Res 1991;A13 (6 & 7):1213-1222.
63. Allen PD, Schimidt TA, Marsh JD, Kjeldsen K. Basic Res Cardiol 1991;87 (Suppl 1):87-94.
64. Dixon IMC, Lee SL, Dhalla NS. Circ Res 1990;66:782-788.
65. Afzal N, Dhalla NS. Am J Physiol 1992;262:H868-H874.

Neuroendocrine Response to Heart Failure

R. Ferrari

Cattedra di Cardiologia, University of Brescia, P.le Spedali Civili 1, 25123 Brescia, ITALY

Summary

The syndrome of congestive heart failure is characterized not only by impaired ventricular function, but also by an increase in some endogenous substances leading to vasoconstriction as well as water and salt retention. Although activation of the systems that release these substances is presumed to be compensatory, the sympathetic nervous system and renin-angiotensin-aldosterone system may contribute to the pathogenesis of the syndrome. Opposite to the effects of these two systems are those evoked by the release of atrial natriuretic peptide which exerts a potent direct vasodilation and natriuresis. In addition, atrial natriuretic peptide inhibits the release of norepinephrine from nerve terminals and suppresses the formation of renin. However, the natriuretic and vasodilator effects of the peptide in patients with congestive heart failure are outweighed by the sodium retention and vasoconstriction caused by sympathetic stimulation and activation of the renin-angiotensin-aldosterone system. The reasons for this are not entirely known. The atrial stretch receptors that are responsible for the release of the peptide become impaired, and it has been suggested that patients with heart failure may adapt to the physiological effects of atrial natriuretic peptide.

Introduction

When the capacity of the heart to maintain cardiac output is impaired, arterial pressure is threatened. As a result, a number of neuro-renal-endocrine mechanisms are activated to preserve circulatory homeostasis and to maintain arterial pressure (1,2). These include activation of the sympathetic and the renin-angiotensin-aldosterone system, and elevated plasma concentration of antidiuretic hormone (vasopressin). Although originally viewed as a beneficial compensatory response (3), the abnormal activation of the renal, nervous, and endocrine mechanisms through which arterial pressure is maintained contributes importantly to the symptoms of heart failure, as well as to the high mortality in this disorder (4,5).

The majority of the studies of the neuroendocrine response to heart failure have been carried out in patients already being treated with diuretics, vasodilators, or angiotensin-converting enzyme inhibitors. Since all these drugs have profound effects on either the sympathetic or on the renin-angiotensin system, it is possible that at least part of the response attributed to heart failure is in reality a response to treatment. We have therefore studied untreated patients with severe heart failure referred to the Postgraduate Medical Institute in Chandigarh (India) and the data on hemodynamics, body water and sodium, renal function, and plasma hormone have been reported before (6-8).

Catecholamines

Many studies have confirmed the original observation of Chidsey et al (9) indicating an increased plasma concentration of norepinephrine in patients with congestive cardiac failure at rest (10-14). The daily urinary excretion of norepinephrine is also increased (15). In addition, the failing myocardium becomes depleted of norepinephrine (16,17). The reasons for this have been debated (18), and a defect in the synthesis and uptake of norepinephrine has been suggested (19). We have shown in experimental animals that the diminution in the total myocardial norepinephrine content is specifically associated with congestive heart failure. While in the presence of ventricular hypertrophy without failure, the content of norepinephrine remains unchanged; its concentration, however, is decreased because the volume of the myocytes is increased (20).

In our untreated patients with cardiac edema, the plasma concentration of norepinephrine was consistently increased, the average being over six times that of the controls. In contrast, the plasma level of epinephrine was unchanged. This is in contrast with the data of Moskowitz et al (21), who showed that in treated patients plasma epinephrine is increased. The increase in plasma norepinephrine provides an indication of the level of activity of adrenergic nerve terminals and suggests that in patients with heart failure, either at rest or during exercise, there is an excessive activation of the sympathetic system (22).

The magnification of the sympathetic outflow is likely to be mediated by the arterial stretch receptors (23). These receptors normally send impulses to the central nervous system that are inhibitory to the activation of two vasoconstrictor mechanisms: the sympathetic nervous system and the release of vasopressin from the pituitary. When these inhibitory afferent impulses are reduced by a fall in systemic arterial pressure, central tonic inhibition is reduced, thereby triggering the activation of sympathetic activity and the release of neurohormones (24). An alternative hypothesis to link a diminished cardiac output with a stimulation of the sympathetic system is that low cardiac output causes underperfusion of the tissues that release an intermediary mediatory substance into the blood. There are, however, several observations that go against this hypothesis (23). The increased sympathetic activity causes tachycardia, improves myocardial contractility, promotes venous constriction, and redistributes arterial resistance. All these mechanisms tend to maintain the systemic arterial pressure.

Whether the effects of the sympathetic nervous system are, as a whole, beneficial is not certain. Vasodilators (25-27) and phentolamine (28,29) clinically improve patients with heart failure. On the contrary, stimulation of mechanical performance with beta-adrenergic agonists has been an unsuccessful approach to the treatment of heart failure (24,30). It should, however, be pointed out here that most oral beta agonists have a vasodilator action that might be beneficial, and dopamine induces renal vasodilation and stimulates diuresis, which also might be clinically useful (33). In addition, most beta agonists have dilatory effects, so they are "inodilators" (34,35) and therefore are able to achieve a positive inotropic effect without any substantial increase in myocardial oxygen uptake. Paradoxically, beta-adrenergic blocking drugs have produced favorable effects in some patients with chronic heart failure (26-28), although serious doubt about their clinical usefulness still exists (27). Consequently the activation of this powerful system is not necessarily beneficial for patients in heart failure. Several factors have to be considered in this respect.

First, in heart failure, there are abnormalities of central neurohormonal regulation. Normally there is a feedback regulation between the sympathetic nervous system and blood pressure as well as intravascular volume. The increased sympathetic activity tends to enhance blood pressure as well as intravascular volume, which, in turn, acting on the high-pressure arterial and atrial baroreceptors, restores the tonic inhibition of neurohormonal activity and thus suppresses the sympathetic outflow. In congestive heart failure, however, the ability of atrial and arterial baroreceptors to suppress sympathetic activity is markedly impaired (36-38). Atrial receptors are no longer activated properly by the increase in atrial pressure. The cause of this desensitization of baroreceptors is not known, but it shifts the balance of circulatory homeostatic mechanisms to an equilibrium in which high circulating levels of norepinephrine and, therefore, vasoconstrictor force are continuously dominant. This leads to a second problem where the ability of the myocardium to respond to continuously high levels of endogenous or exogenous catecholamines becomes attenuated (39-42). Recent work on samples of heart muscle obtained from patients undergoing heart transplantation has shown that beta receptors are reduced in number, stimulation of adenyl cyclase by isoprenaline is diminished, and the contractile response to catecholamines is lessened (43,44). These findings have generated the idea that beta receptors in heart failure are downregulated. The consequence would be that patients with heart failure will be less responsive to adrenergic stimuli than normal persons, and the positive effects to sympathetic stimulation would diminish with time as the receptors are downregulated. Of course, this hypothesis may be oversimplistic. Heart muscle from patients with severe heart failure also seems to be less responsive to other drugs (45,46). Since the myocardium contains both beta $_1$ and beta $_2$ receptors (47) it is not clear which receptors are downregulated and what specific defect at the molecular level causes catecholamines resistance in heart failure (48). Clearly, further studies are needed.

Nevertheless, it is important to emphasize here that the reduction in catecholamine responsiveness is relatively selective for the myocardium. Catecholamines do not become depleted in the kidney (19), where downregulation does not take place. Furthermore, sympathetic neurotransmitter activity is augmented, rather than attenuated, in the peripheral blood vessels (49). This explains why catecholamines in the long term are not necessarily beneficial for the patient with heart failure. Increased sympathetic activity will support excessive vasconstriction with increased ventricular afterload and, through the stimulation of the renin-angiotensin-aldosterone system, it leads to water retention and subsequent increased ventricular preload. This will impose a greater load of work on the myocardium, which cannot benefit from the positive inotropic effect mediated by catecholamines. In turn, this will cause further deterioration of the cardiac pump, worsen the heart failure, activate neuroendocrine systems, and further increase the systemic resistance. This would represent, as suggested (5), a progressive spiral of heart failure.

Renin-Angiotensin-Aldosterone

Early measurements of both aldosterone and renin activity in the plasma of patients with heart failure were confusing. Brown et al (50) found plasma renin to be increased only in one third of patients with untreated congestive cardiac failure from different causes. Recently, Bayllis et al (51) and Kubo et al (52) reported normal values for plasma renin activity in untreated patients with clinically mild heart failure. The same was true for aldosterone (51). This is quite in contrast with the common belief that the renin-angiotensin-aldosterone system is stimulated in heart failure (53-55). However, no meaningful conclusion can be made from these studies because plasma renin activity and aldosterone concentrations are known to be influenced by diuretic therapy (51,56).

In our untreated patients with cardiac edema and, therefore, an excess accumulation of water, plasma renin activity and aldosterone levels varied a great deal. Three patients had normal values, while the remaining five had very high values. It seems, therefore, that both aldosterone and renin activity are increased in some patients independently from administration of diuretic therapy, but are normal in others, despite the presence of water retention. An explanation for these results is provided by the experiments on dogs described by Watkins et al (57). They showed that after constriction of the pulmonary artery or inferior vena cava, the initial response was a reduction in blood pressure and a rise in plasma renin activity, plasma aldosterone and water intake. However, renin activity and aldosterone returned to normal levels in the following days, as soon as the plasma volume expanded and the arterial blood pressure was restored. In those animals in which blood pressure was not restored, plasma renin activity and plasma aldosterone remained elevated throughout the period of constriction, suggesting the existence of a feedback mechanism between plasma volume and renin activity. Thus, it is likely that the activation of this system is transient rather than sustained and probably is related to different phases during the development of heart failure (24). It has been suggested that plasma renin activity and angiotensin II are high in the initial stages of heart failure. In mild congestive chronic failure, they tend to fall toward normal as the condition is established whereas in severe heart failure they remain high (37).

The increased release of renin from the juxtaglomerular cells has been found to depend on changes in cardiac output that threaten the renal blood flow and arterial perfusion pressure (24,58). In addition, changes in the concentration of sodium ions in the region of the macula densa affect the rate of release of renin (59). The responsiveness of the system is greatly increased by adrenergic stimuli, which act by enhancing the threshold for renin release (59).

The increased concentration of angiotensin II that will ultimately result from the increased plasma renin activity has other important properties in heart failure in addition to stimulating the secretion of aldosterone. Angiotensin II acts on the efferent arterioles to increase the glomerular hydraulic filtration pressure (60); it is a powerful vasconstrictor, it enhances neuroeffector transmission, and it is concerned with the sensation of thirst.

Atrial natriuretic peptide is able to directly suppress renin release (37), and, in addition, it exerts effects that are opposite to the renin-angiotensin-aldosterone system. However, as previously discussed, the function of arterial and atrial baroreceptors is likely to be impaired in congestive heart failure. Thus, atrial distension does not suppress sympathetic activity in this disorder (61), and it has been shown that angiotensin II can directly attenuate barocreceptor sensitivity (36).

The stimulation of the renin-angiotensin system by the low arterial pressure, leading to a retention of water by the kidneys, may be seen as a beneficial mechanism in increasing the filling of the heart and, thereby, its output. However, a mainstay of the drug treatment for heart failure is the optimal use of diuretics (5). In addition, long-term treatment with angiotensin-converting enzyme inhibitors has been shown in several trials to be beneficial for patients in heart failure (23,62-65). The benefit is not simply due to a vasodilator action. These drugs probably inhibit the facilitatory action of angiotensin II on norepinephrine and, indirectly may enhance baroreceptor sensitivity (36). Interestingly, this group of drugs, which appears at present to be the most powerful in the treatment of heart failure, can be seen to produce good results when used with diuretics (66). Thus, the converting-enzyme inhibitors may be regarded as preventing the consequences of some undesirable side effect of diuretics, which stimulate the renin-angiotensin system. Similarly, the effects of aldosterone could be beneficial, but spironolactone is found to be helpful in the therapy of congestive cardiac failure (58). Thus, as in the case of catecholamines, it seems that every time we are counteracting the effects of the renin-angiotensin-aldosterone system, the patients improve.

Vasopressin

It has been reported that an increased concentration of vasopressin may occur in patients with severe congestive heart failure and after forced diuresis (67-70). However, recent investigations in treated patients with congestive heart failure have given diverse results. Perry and Fyles (71) found no difference between patients in congestive heart failure and normals, while others (72-74) found vasopressin to be increased in only some patients. In our untreated patients with cardiac edema, the plasma concentration of vasopressin in general was not increased. Only in one case was it abnormally high. This is not entirely surprising if we consider that hyponatremia and consequent hypoosmolarity of the plasma is usual in congestive heart failure (73-75), so that there is no osmotic stimulus to the release of vasopressin. On the other hand, nonosmotic stimuli may be operating through the baroreceptors to increase the release of vasopressin from the pituitary. Opposing this mechanism, the distension of the atria would be expected to decrease vasopressin release, but, once again, in patients with heart failure reflexes mediated by atrial distension are not operating properly (75). Thus, from our study and other studies, it appears that vasopressin is not necessarily increased in untreated patients with cardiac edema, but it is probably increased with respect to the osmotic pressure of the plasma.

Atrial Natriuretic Peptide

Microscopic studies had shown the presence of specific granules in atrial myocardium, the number of which varied with changes in the water and electrolyte balance (75). In addition, De Bold (76) demonstrated that an atrial extract has a natriuretic effect; the active principle has now been identified as a polypeptide and its amino acid sequence has been determined (77,78). Increased concentrations of atrial natriuretic peptide have been found in patients with treated congestive heart failure (79-89). In our untreated patients with cardiac edema, the plasma concentration of atrial natriuretic peptide was dramatically raised in every case, the average being 14 times the controls.

The release of the peptide is stimulated by stretching of the atrial tissue (90) and by volume loading of the circulation (78). The increased concentration in the blood of patients with congestive heart failure is due to excessive release from overstretched atria. Once released, atrial natriuretic peptide exerts a potent direct vasodilation, probably by increasing the intracellular cyclic GMP and natriuresis. In addition, under some experimental conditions, atrial natriuretic peptide inhibits the release of norepinephrine from nerve terminals, as well as the vasoconstrictor action of norepinephrine on systemic vessels (91).

Atrial natriuretic peptide also suppresses the formation of renin and counteracts the systemic vasoconstrictor actions of angiotensin II, as well as the ability of angiotensin II to stimulate thirst and to stimulate the secretion of aldosterone (92). All these mechanisms, therefore, act together to lower the systemic vascular resistance and to increase the excretion of sodium. This will unload the heart and reduce its energy consumption. However, clearly, the natriuretic and vasodilator effects of the peptide in patients with congestive heart failure are outweighed by the sodium retention and vasoconstriction caused by sympathetic stimulation and activation of the renin-angiotensin-aldosterone system. The reasons for this are not known at present. As already mentioned, the atrial stretch receptors become impaired in chronic heart failure. Consequently, the ability of atrial distension to increase the release of atrial natriuretic peptide is blunted (37), and the slope of the curve of atrial-pressure/atrial-natriuretic-peptide release is shifted for that circulating level of the atrial natriuretic peptide for a given atrial pressure is reduced (37). In addition, it has been suggested that patients with heart failure may adapt to the physiologic effects of atrial natriuretic peptide over time (24), and in particular the action of the peptide on the kidney is markedly attenuated (93).

Conclusions

Originally the neuroendocrine response in congestive heart

failure was viewed as a specific compensatory beneficial mechanism. Over the past years, however, we have learned that this view is too simplistic. All the changes in the peripheral autonomic and in the renin-angiotensin-aldosterone systems reviewed above, exacerbate the condition of heart failure. Diminution of baroreceptor response reduces the inhibitory control that is exerted on the vasomotor centers.

Decreased sensitivity of atrial receptors decreases the afferent input through which atrial distension reduces sympathetic outflow. A lack of responsiveness of the myocytes to postganglionic sympathetic stimulation reduces the increase in output necessary to maintain blood pressure. This, in turn, stimulates further the sympathetic and renal response, leading to deleterious vasoconstriction and retention of salt as well as water, which is clinically evident as edema. Consequently pharmacologic treatment aimed at reducing the effects of the neuroendocrine response has proven to be advantageous for patients.

The problem is that the response of the body to heart failure is not specific, as was originally thought (1,2,58). It has been suggested (1,2,58) that what is evoked is a stereotyped neuroendocrine response for which we have been programmed by natural selection in order to maintain arterial pressure. Therefore, for our survival in life threatening circumstances such as hypovolemia, trauma and hemorrhage, as well as in exercise, on which daily life depends, the neuroendocrine response is very useful for the short-term. Heart failure, however, is often a chronic state in which the arterial pressure is continuously and chronically threatened due to a primary abnormality in the cardiac pump. Because of that, the neuroendocrine response through which arterial pressure should be maintained, intensifies to an abnormal degree and a vicious circle is established.

References

1. Harris P. Lancet 1988;1036-1038.
2. Harris P. Cardiovasc Res 1983;313:373-437.
3. Braunwald E. In: Heart Disease. A Textbook of Cardiovascular

Medicine (ed.) E Braunwald. WB Saunders Co., Philadelphia.
4. Packer M, Lee WH, Kessler PD. Circulation 1987;75:IV-80.
5. Poole-Wilson PA. Med Interne 1985;2:866-871.
6. Anand IS, Ferrari R, Kalra GS, Wahi PL, Poole-Wilson PA, Harris PC. Circulation 1989;80:209-305.
7. Anand IS, Ferrari R, Kalra GS, Wahi PL, Poole-Wilson PA, Harris P. Circulation 1991;83:1880-1887.
8. Ferrari R, Anand I. Cardiovasc Drugs Ther 1989;3:979-986.
9. Chidsey CA, Harrison DC, Braunwald E. N. Engl J Med 1962;267:650-654.
10. Goldstein DS. Am J Cardiol 1981;48:1147-1152.
11. Francis GS, Cohn JN. Ann Rev Med 1986;37:235-247.
12. Francis GS, Rector TS, Cohn JN. Am Heart J 1988;116:1464-1468.
13. Francis GS, Goldsmith SR, Pierpont G, Cohn JN. J Lab Clin Med 1984;103:393-398.
14. Francis GS. Cardiovasc Res Rep 6:444-454.
15. Rutenberg HL, Spann JF Jr. In: Congestive Heart Failure (ed.) DT Mason. 1976; York Medical Books, New York.
16. Chidsey CA, Braunwald E, Morrow AG. Ann J Med 1965;39:442-451.
17. Chidsey CA, Sonnenblick EH, Morrow AG, Braunwald E. Circulation 1966;33:43-51.
18. Rupp H, Jacob R. In: The Regulation of Heart Function (ed.) H Rupp. Thieme, New York, 1986; pp. 53-70.
19. Spean JF, Chidsey CA, Pool PE, Braunwald E. Circ Res 1965;17:312-321.
20. Ceconi C, Condorelli E, Rodella A. International Symposium: Physiology Clinical Aspects and treatment of Coronary Insufficiency. Torino, 3-6 June, 1986; pp. 16.
21. Moskowitz RM, Kinney EL, Zells RF. Chest 1979;76:640-646.
22. Lake CR, Ziegler MG, Kopin IJ. Life Sci 1976;18:1315-1321.
23. Harris P. Br Heart J 1987;58:190-203.
24. Packer M. Circulation 1988;77:721-730.
25. Harper RW, Claxton CH, Middlebrouk K. Med J Aust 1980;II:36-38.
26. Francinosa JA, Goldsmith SR, Cohn JN. Am J Cardiol 1980;69:559-566.
27. Lipkin DP, Poole-Wilson PA. Br Med 1985;291:993-996.

28. Gould L, Zahir M, Ettinger S. Br Heart J 1969;31:154-157.
29. Williams DO, Hilard GK, Cantor SA. Clin Res 1975;23:88-93.
30. Heilbraun SM, Shah P, Valentine HA. Circulation (abstract) 1986;74:II-310.
31. Fowler MB, Bristow MR. Am J Cardiol 1985;55:120D-124D.
32. Swedberg K, Waagstein F, Hjarlmarson A, Wallentin I. Lancet 1979;1:1374-1376.
33. Katz AM. N Engl J Med 1978;299:1409-1410.
34. Opie LH. Lancet 1986;7:1336.
35. Kumada T, Kawai C. Cardiovasc Drugs Ther 1989;2:751-755.
36. Hirsch AT, Dzau VJ, Creager MA. Circulation 1987;75:IV-36.
37. Edwards BS, Miller WL, McGopn MD, Burnett JC. J Am Coll Cardiol (abstract) 1988.
38. Packer M, Lee WH, Medina N, Yushak M. Circulation 1987;76:IV-357.
39. Covell JW, Chidsey CA, Braunwald E. Circ Res 1966;19:51-72.
40. Feldman MD, Copelas L, Cwathmey JK. Circulation 1987;65:331-343.
41. Fowler MB, Laser JA, Hopkins GL. Circulation 1986;74:1290-1299.
42. Vatner DE, Vatner SF, Kujii AM, Homcy CJ. J Clin Invest 1985;76:2259-2264.
43. Bristow MR, Ginsburg R, Minobe W. N. Engl J Med 1982;307:205-211.
44. Bristow MR, Kantrowitz NE, Ginburg R, Fowler MB. J Mol Cell Cardiol 1985;17:41-52.
45. Brown L, Lorenz B, Erdmann E. Cardiovasc Res 1986;20:516-520.
46. Wilmshurst PT, Walker JM, Fry CH. Cardiovasc Res 1984;18:302-309.
47. Bristow MR, Ginsburg R, Umans V. Circ Res 1986;59:297-309.
48. Longbaugh JP, Vatner DE, Jujii AM. Circulation (abstract) 1986;74:II-198.
49. Kramer PS, Mason DT, Braunwald E. Circulation 1968;38:629-639.
50. Brown JJ, Davies DL, Johnson VW. Am Heart J 1970;80:329-342.
51. Bayliss J, Norell M, Canepa Anson R. Br Heart J 1987;57:17-22.
52. Kubo SI, Clark M, Laragh JH. Am J Cardiol 1987;60:1322-1328.
53. Francis GS. Am J Cardiol 1985;21A.

54. Francis GS, Goldsmith SR, Levine TB. Ann Intern Med 1984;101:370-377.
55. Riegger AJG. Eur Heart J 1985;6:479-489.
56. Nicholls MG, Espiner EA, Donald RA, Hughes H. Clin Sci Mol Med 1974;47:301-315.
57. Watlins L Jr., Burton JA, Haber E. J Clin Invest 1976;57:1606-1607.
58. Harris P. Eur Heart J 1982;3:5-10.
59. Skinner SL, McCubbin JW, Page IH. Circ Res 1964;15:64-76.
60. Packer M, Lee WH, Kessler PD. Circulation 1986;74:766-779.
61. Zucker IH, Share L, Gilmore JP. Am J Physiol 1979;236:H554-H567.
62. Cleland JGF, Dargie HJ, Hodsman GP. Br Heart J 1984;52:530-535.
63. Kramer BL, Massie BM, Topic N. Circulation 1983;67:807-816.
64. Sharpe DN, Murphy J, Coxen R, Hannan SF. Circulation 1984;70:271-278.
65. Franciosa JA, Wilen MM, Jordon RA. J Am Coll Cardiol 1985;5:101-107.
66. Parcker M. Int J Cardiol 1985;7:111-120.
67. LeJemtel TH. J Am Coll Cardiol 1986;7:766-767.
68. Creager MA, Faxon DP, Cutler SS. J Am Coll Cardiol 1986;7:758-765.
69. Cannon PJ. N Engl J Med 1977;296:26-32.
70. Yamane Y. Jpn Circ J 1968;32:745.
71. Perry WF, Fyles TW. J Clin Endocrinol Metab 1953;13:64-72.
72. Stein M. Schwartz R, Mersky IA. J Clin Invest 1954;33:77-81.
73. Szatalowicz VL, Arnold PE, Chaimovitz C. N Engl J Med 1981;305:263.
74. Yamane Y. Jpn Circ J 1968;32:745-759.
75. Riegger GAJ, Liebau G, Kochsiek K. Am J Med 1982;72:49-61.
76. de Bold AJ. Proc Soc Exp Biol Med 1979;161:508-511.
77. Atlas SA, Kleinert HD, Camargo MJ. Nature 1984;309:717-719.
78. de Bold AJ, Borenstein HB, Veress AT, Sonnenberg H. Life Sci 1981;28:89-94.
79. Hartter E, Weissel M, Stummvoll HK. Lancet 1985;2:93-94.
80. Lang RE, Tholken II, Genten D. Nature 1985;314:264-266.

81. Patterson A, Hedner J, Hedner T. Eur Heart J 1986;7:693-696.
82. Richards AM, Cleland JGF, Tonolo G. Br Med J 1986;293:409-412.
83. Shenker Y, Sider RS, Ostafin EA, Grekin RJ. J Clin Invest 1985;76:1684-1687.
84. Riegger GAJ, Kromer EP, Kochsiek KJ. Cardiovasc Pharmacol 1986;8:1107-1112.
85. Schriffin EL. N Engl J Med 1986;315:765-766.
86. Tikkanen I, Fyrquist F, Metsarinne K, Leidenius R. Lancet 1985;2:66-69.
87. Raine A, Erne P, Burgisser E. N Engl J Med 1986;315:533-537.
88. Arendt RM, Gerbers AL, Ritter DJ. Hypertension 1984;4:S131-S135.
89. Ogawa K, Ito T, Hashimoto H. Lancet 1986;1:106-109.
90. Agnoletti G, Rodella A, Ferrari R, Harris P. J Mol Cell Cardiol 1987;19:217-220.
91. Mark AL. Fed Proc 1987;46:36.
92. Laragh JH. N Engl J Med 1985;313:1330-1343.
93. Shaknovitch A, Pondolfino K, Clark M. J Clin Invest 1986;78:1362-1373.

Alterations of Extracellular Matrix in Cardiac Atrophy

B.V. Shekonine[1], B. Korecky

Department of Physiology, Faculty of Medicine,
University of Ottawa, Ottawa, Ontario CANADA K1H 8M5

Introduction

After the introduction of microvascular sutures in the 60's techniques for the transplantation of many organs were developed in the next 10 years in small laboratory animals. Recently several investigators used heterotopically transplanted hearts in inbred rats as a useful model for functional and morphological studies of the whole heart (for brief review see 1). This transplant functions as a denervated non-working Langendorff heart with preserved coronary flow, beats mostly isovolumically and does not appreciably contribute to the hemodynamic of the recipient. The recipient heart seems to be unaffected by the surgical procedure and the presence of the transplant and consequently maybe used as a control heart that fully supports hemodynamic load essential for the recipient organism (2).

A considerable atrophy of the left ventricle develops in time which may be explained primarily by atrophy of individual myocytes (3). In addition, visible relative changes in extracellular matrix occur which are due initially to a relatively slow turnover of the extracellular matrix components (e.g. collagen) as compared to contractile proteins leading to formation of interstitial fibrosis which may have either a reparative character due to perioperative ischemia or a reactive one as a response to alteration of hemodynamics of the transplant or to its heterotopic location.

We studied the onset and later development of alterations of the

[1]*Cardiology Research Center, Moscow 121551 USSR*

extracellular matrix in this model using histological and immunohistochemical techniques. We were able to visualize collagen type I and fibronectins in the heterotopically isotransplanted rat heart located in the abdominal cavity and compare their deposition to the recipient heart in situ and to the in situ hearts of normal and sham operated animals.

Materials and Methods

Experimental Animals

Hearts of Fisher inbred male rats (Charles River Laboratories, Inc., Wilmington, Mass.) weighing 330-350 g body wt were transplanted into the abdominal cavity of recipients of the same strain by using a modified technique of Ono and Lindsey, as described in detail by us in 1983 (2). In the present experiments, the donor heart was exposed to cold (3-5°C) cardioplegia using a modified solution (Plegisol, Abbott Laboratories, North Chicago) with a potassium concentration of 16 mM. The donor heart received 15-25 ml of the above solution by retrograde perfusion through the aortic stump with pressures not exceeding 50 mm Hg. After recovery, the rats were put on a standard diet and received antibiotics for 7 days (Gentocin 1.5 mg/day). The beating of the transplants in both groups was monitored by palpation.

Four experimental groups were selected for further investigation at different time periods after transplantation: an early post-operative period when atrophy develops, a subsequent period of stable atrophy, a period of stable atrophy with probable onset of fibrosis and a period of long term atrophy with stabilized degree of fibrosis.

The recipient rats were then anesthetized, their abdomen opened, the transplant isolated, inspected and subsequently the animal was bled. Both the recipient in situ heart and the transplanted heart were removed, the atria trimmed off, the remaining ventricles plus septum were washed in saline, blotted dry and weighed. Then the base of the heart was trimmed off, a part of the middle section put in neutral phormol and used for histology while the other part was frozen in liquid nitrogen and used for immunochemistry. The apex was left to determine the dry weight.

Polyclonal Antibodies

Antiserum against fibronectin was obtained by immunizing rabbits with 0.5 mg fibronectin isolated from human plasma by affinity chromatography on gelatin-Sepharose and further purified on DEAE-cellulose as described previously (4). In our experiments the antigen was injected in Freund's complete adjuvant 3-4 times intracutaneously with 3 week intervals between injections. A coarse fraction of immunoglobulins was precipitated by a 50% ammonium sulphate solution. Polyclonal antibodies against fibronectin were isolated from the immunoglobulin fraction by affinity chromatography on fibronectin-Sepharose and used for immunofluorescence at a concentration of 0.025 mg/ml.

Polyclonal antibodies against rat type I collagen were kindly provided by Dr. A.E. Berman (Institute of Medical and Biological Chemistry, Academy of Medical Sciences, Moscow, USSR). Type I collagen was isolated from rat tail tendons by acetic acid extraction and subsequent precipitation in acid and neutral NaCl solutions. It was analyzed for purity with SDS-PAGE in 10% polyacrylamide. Rabbits were immunized subcutaneously with 0.5 mg of collagen with Freund's adjuvant (Gibco) 3-4 times at 3-week intervals. The prepared antiserum was applied to an immunoabsorbent chromatography column prepared by immobilization of collagen to CNBr-activated Sepharose 4B (Pharmacia Fine Chemicals). Eluted antibodies are specific as shown after testing by indirect ELISA using peroxidase-labelled goat antibodies to rabbit immunoglobulins. The above technique is described in more detail (5). All polyclonal antibodies which we used in this study were monospecific and reacted only with the appropriate antigen in immunoblotting experiments.

Monoclonal Antibodies

Mouse monoclonal antibody MonAB IST-9 which recognizes the ED-A sequence present only in cellular fibronectin (6,7) was a generous gift of Dr. L. Zardi, (Institute Nazionale per la Ricerca sul Cancro, Genoa, Italy). Undiluted culture supernatants were used for immunofluorescence. Mouse MonAB Lg was kindly provided by Dr. O. Printseva (Institute of Experimental Cardiology, Moscow, USSR). This antibody reacted with all forms of fibronectin in rat and human tissues. It was produced in an identical way as the L_8 form which is described in detail elsewhere (8). For immunocytochemical analysis ascitic fluid was diluted 1:100.

Immunofluorescence

Serial unfixed cryostat sections 5 micrometers thick were incubated with primary antibodies for one hour at room temperature, washed in phosphate buffered saline (PBS) at pH 7.4 for 30 minutes. Subsequently they were incubated with biotinylated sheep antimouse IgG or donkey antirabbit IgG (Amersham 1:100 dilution) for 45 minutes. After washing in PBS, sections were incubated with Texas Red streptavidin (Amersham 1:100 dilution) for 30 minutes, thoroughly washed, air-dried and mounted in UVinert embedding medium (Serva). Control stainings were performed by replacing the primary antibody with 1gG fraction from non-immunized rabbit or mouse blood serum. The control stainings were either negative or showed non specific reaction only. Several cryostat sections from each experimental series were fixed in formalin and stained with hematoxylin and eosin. Specimens were observed with a Zeiss Axiophot microscope, photographs of the free wall of the left ventricle were taken with Kodak TMAX-400 film. The deposition of collagen I and fibronectin as manifested by the intensity of fluorescence and the width of the intracellular space in each transplant was assessed subjectively by one of us (B.S.) and compared to the respective recipient hearts. The data were then combined within each group and ranked from 0 to +++ in the subendocardium and midmyocardium.

Results

Myocardium of sham operated and recipient rats

We observed that the distribution of collagen type I in the left ventricular myocardium of normal rats, sham operated rats and in the in situ hearts of recipients with viable transplants was the same as previously described in normal human and mouse hearts (9,10). It was predominantly abundant in the extracellular matrix (ECM) of the

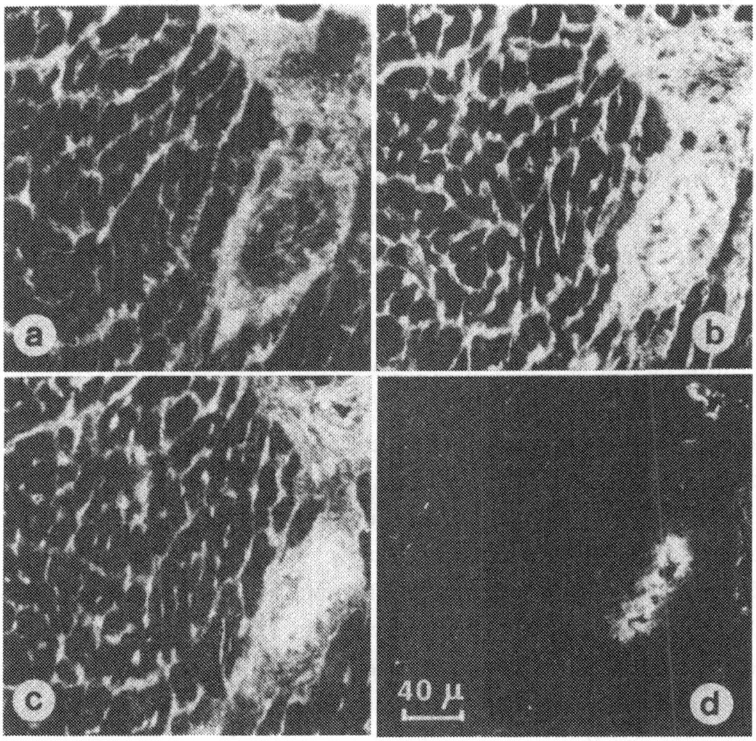

Fig. 1: Immunofluorescent staining of collagen type I and fibronectin in serial sections of the myocardium of a sham operated rat 14 days after surgery: (a) identification of type I collagen; (b)-(d) visualization of fibronectin in the same region of the left ventricle: (b) identification of fibronectin with polyclonal antibodies, (c) with monAB Lg against different mammalian fibronectins, (d) with monAB IST-9 against ED-A sequence. x250

endocardium and epicardium. However, in the midmyocardium it was present to a lesser degree between individual cardiomyocytes but was prominent in the connective tissue around small and large vessels (Fig 1a). The perimysial strands of connective tissue around the groups of cardiomyocytes appeared especially rich in collagen type I. In the hearts of sham operated rats and in the in situ hearts of the recipients with transplants in the abdomen, we observed that the localization of fibronectin was the same as in normal rats reported previously (11).

The presence of fibronectin was detected with polyclonal

antibodies and MonAB Lg which recognized different forms of mammalian fibronectin. This glycoprotein was visualized in endocardium, and epicardium both in and around blood vessel walls and between the cardiomyocytes (Fig 1b,c). Immunocytochemical staining with MonAB IST-9 against ED-A sequence revealed that the cellular form of fibronectin was predominantly located under the endothelium of blood vessels and in the endocardium. Weak or no immunofluorescence was observed around cardiomyocytes (Fig. 1d).

Transplants 3-17 days old

In three out of a total of fifteen rats which we analyzed in this group we observed transmural infarctions in their left ventricles and these hearts were excluded. In the remaining hearts mostly normal myocardium was seen during detailed histological examination. However, in addition to normal myocardium occasional small patches of granulation tissue were seen in only 5 out of the 12 analyzed transplants. They were predominantly located in the subendocardium (Fig. 2a) and were due to perioperative ischemia. They contained fibroblasts, macrophages, polymorphonuclear leucocytes, lymphocytes and plasma cells. The ECM of these patches stained strongly with polyclonal antibodies against type I collagen (Fig. 2b), fibronectin (Fig. 2c) and also with MonAB Lg against an epitope common to different forms of mammalian fibronectin (Fig. 2d). Positive reaction with MonAB IST-9 against ED-A sequence detected local 'de novo' synthesis of fibronectin (Fig. 2e). For control staining see Fig. 2f.

In addition to the above patches, in all transplants a mild diffuse fibrosis was also detected which could be classified as a reactive fibrosis. In this latter case we observed that the intracellular spaces around the atrophic myocytes showed increased specific immunofluorescence for collagen and fibronectin indicating an increased volume of ECM. This type of fibrosis was initiated in the midmyocardium predominantly near the subendocardial areas and around small blood vessels. In Figure 3 a development of reactive fibrosis is displayed in two 15 day old transplants. The first row (a,b,c) represents a milder stage and the second row (d,e,f) a more

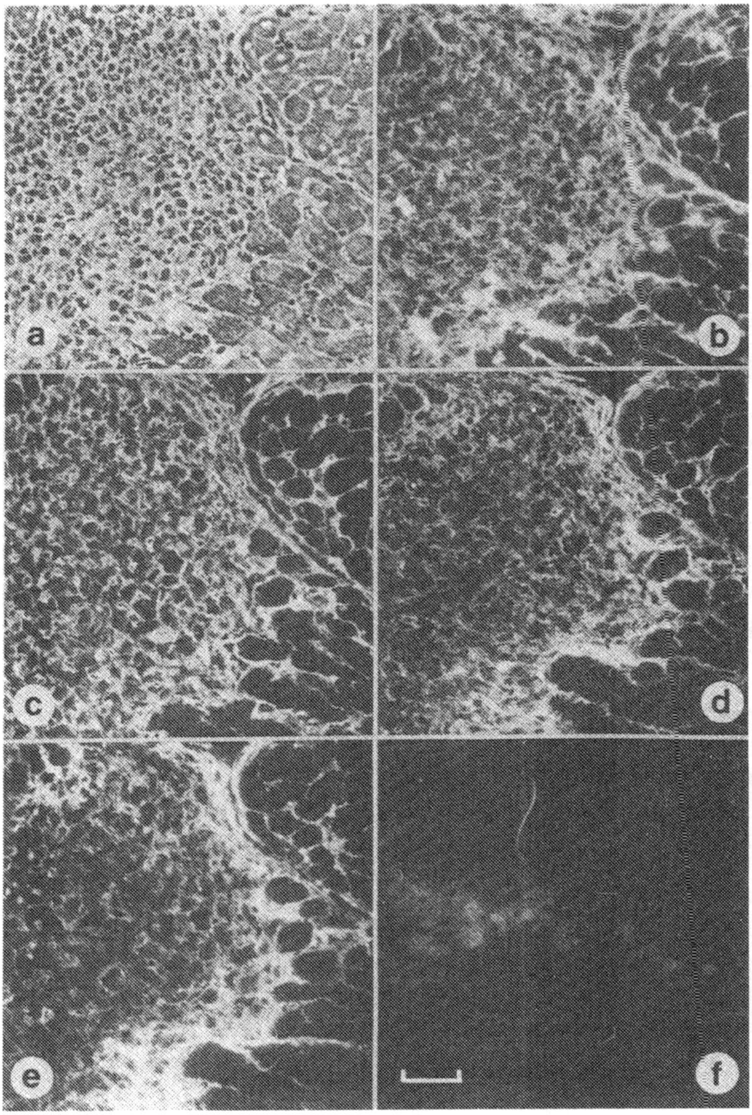

Fig. 2: Granulation tissue in the subendocardium of a three day old transplant (a) hematoxylin and eosin staining; (b)-(f) immunofluorescence on serial sections of the same region: (b) collagen type I; (c) fibronectin revealed with polyclonal antibody (d) fibronectin after incubation with MonAB Lg; (e) cellular fibronectin identified with MonAB IST-9; (f) control staining. x250

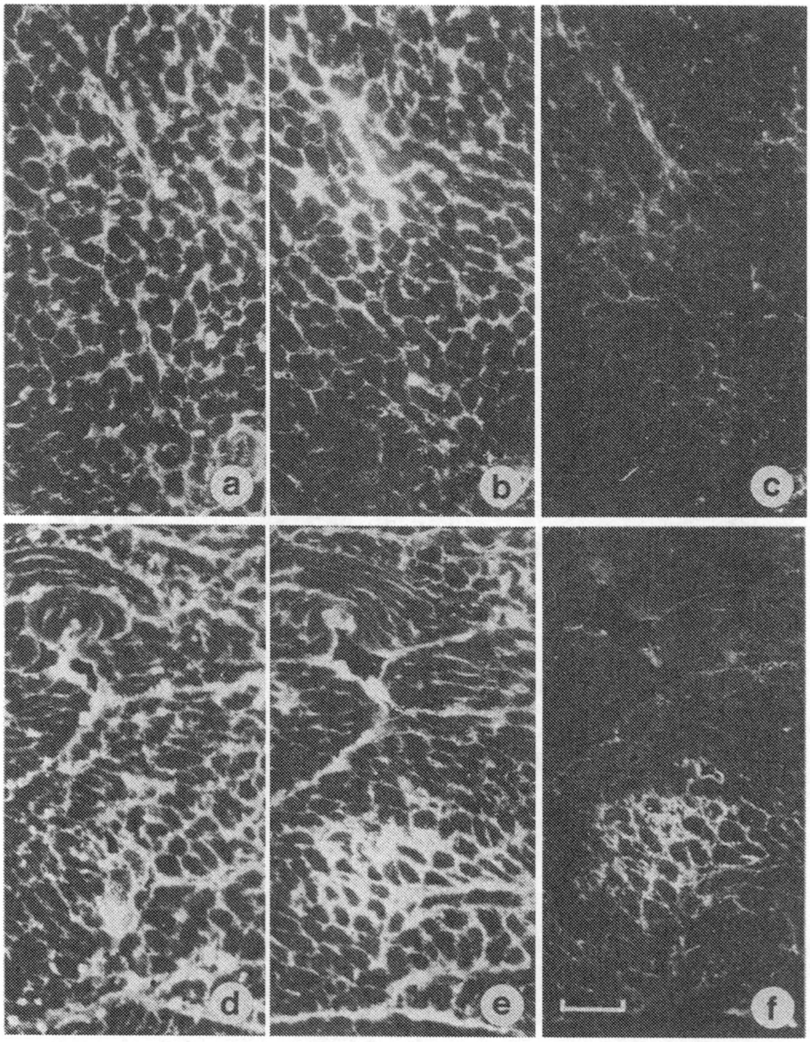

Fig. 3: Increased volume of extracellular matrix in two transplants 15 days old. (a)-(c) Milder stage of reactive fibrosis; (a) immunofluorescence of collagen type I; (b) fibronectin as revealed with MonAB Lg; (c) fibronectin in the intracellular spaces detected with MonAB IST-9. (d)-(f) Progression of diffuse fibrosis: (d) type I collagen; (e) fibronectin identified with MonAB Lg; (f) focal fluorescence of fibronectin detected with MonAB IST-9. x250

advanced stage of reactive fibrosis. When compared to the in situ hearts of normal and sham operated animals in the milder stage we saw no detectable increase in immunofluorescence of collagen around the myocytes (a), no visible change in fluorescence of polyclonal antibodies against total fibronectin (b), whole the fluorescence of cellular fibronectin as displayed by monoclonal antibodies was higher.

As compared to the milder stage of reactive fibrosis in the subsequent more severe stage we saw a considerable increase in fluorescence of collagen (d), total fibronectin (e) and a focal appearance of fluorescence of cellular fibronectin (f) in the ECM around the myocytes. In both hearts subendocardial areas are displayed. Different degree of progression of fibrosis was achieved by the 15th day after transplantation.

Transplants 20-35 days

In five animals with viable transplants which showed normal rate of beating and after laparotomy displayed no visible scars and adhesions, histological examination of the hearts showed the following. In all transplanted hearts signs of atrophy of myocytes in the left ventricles were detected. In the patches of granulation tissue which were observed in some transplants a proliferation of fibroblasts to various degrees could be shown. Some of the patches were being transformed into fibrous scars as shown by the presence of collagen type I and fibronectin by immunohistochemical staining.

Marked atrophy of individual myocytes was accompanied by increased accumulation of fibrous tissue in the ECM, leading to diffuse fibrosis particularly in the subendocardial area of the left ventricle. This was corroborated by increased immunofluorescence of collagen type I and fibronectin as demonstrated by immunohistochemistry in the ECM surrounding individual myocytes. During this phase of developing diffuse fibrosis we could detect fibronectin with polyclonal antibodies and MonAB L g which react with various forms of fibronectin. However, unlike in the earlier stages of the diffuse fibrosis we could not show local presence of fibronectin by using MonAB-IST-9 which reacts specifically with the

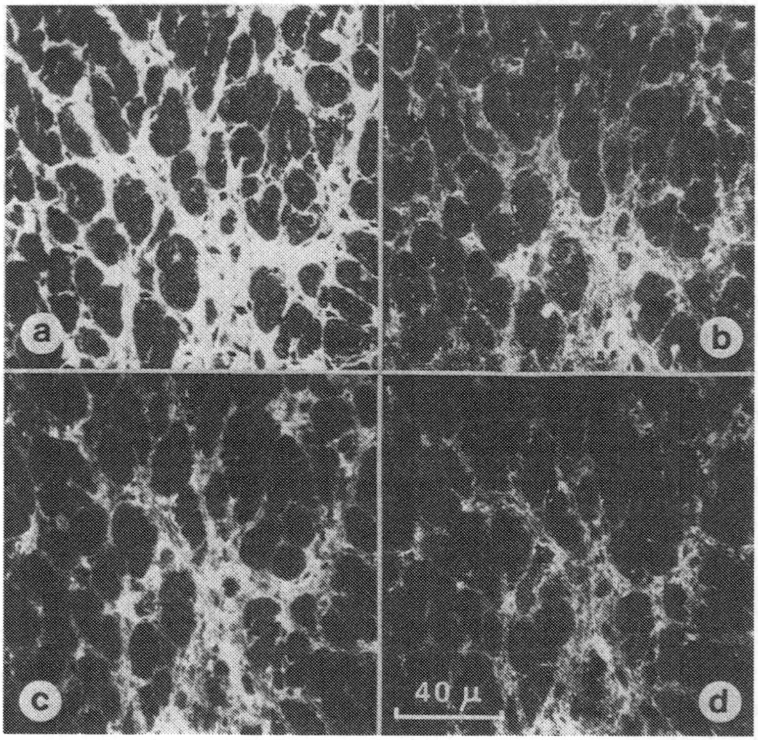

Fig. 4: Pronounced reactive fibrosis in the myocardium of transplanted heart 40 days after the surgery. (a)-(b) Immunofluorescence in serial sections of the left ventricle: (a) high content of collagen type I in the fibrotic intracellular spaces; (b) identification of the fibronectin with polyclonal antibodies; (c) fibronectin revealed with MonAB L 9 ; (d) immunofluorescence with MonAB IST-9. x400

ED-A sequence of cellular fibronectin. Consequently no proof of increased synthesis of cellular fibronectin could be obtained as no fluorescence was detected (data not shown).

Transplants 40-135 days old

After this extended time period subendocardial scars and diffuse reactive fibrosis were found in all transplanted hearts. Diffuse fibrosis was more pronounced in the subendocardial areas of the left

Table 1.

Antibodies	Reparative Fibrosis				Reactive Fibrosis			
	A	B	C	D	A	B	C	D
Anticollagen	+	++	+++	+++	+	++	+++	+++
Poly anti-fibronectin	+++	+++	++	+	++	+++	++	+
MonAB Lg (anti cFn + pFn)	+++	+++	++	+	+	++	+	+
MonAB IST-9 (anti cFn)	+/o	++	+/o	+/o	+/o	++	+/o	+/o

Deposition of collagen I and fibronectins in transplanted hearts as determined by immunohistochemical staining (cFn-cellular fibronectin, pFn-plasmatic fibronectin) in four groups of rats during the following time periods after transplantation. A: 3-5 days, n-4; B: 7-15 days, n-8; C: 25-30 days n-5; D: 40-120 days, n-9. Mass of the transplanted hearts as compared to the respective recipient hearts (100%) expressed as mean value for each group: A-81%, B-69%, C-62%, D-59%.

ventricle and around the blood vessels. However, no signs of rejection of cardiomyocytes were seen. Immunohistochemical examination showed the existence of thick fibres of collagen type I that cross through wide intracellular spaces around the atrophied cardiomyocytes (Fig. 4a). Unlike the situation in the earlier postoperative time period when the detectable amounts of fibronectin was predominant, at this later stage fibronectin was not overtly abundant in the interstitium and could only be detected in smaller quantities with polyclonal antibodies and two types of MonAB (Fig. 4b,c,d).

Discussion

The results of our studies indicate that during the development of experimental left ventricular atrophy in isotransplanted hearts a gradual relative increase in the apparent size of extracellular matrix between the myocytes and an enhanced fibrosis may be observed. Within the first five to ten days this is primarily due to the fact that the myocytes atrophy quite considerably (3) while the volume of extracellular matrix probably remains relatively unchanged. In addition, a new synthesis of fibrous material in the extracellular matrix could be detected at this time. The data are summarized in Table 1.

Two time dependent types of fibrosis quite different in etiology and morphogenesis could be demonstrated in the transplant. First the granulation tissue at the site of postoperative necrosis of cardiomyocytes led to the development of scars (i.e. reparative fibrosis). The second, later developing type of fibrosis, was detected as a diffuse increase in the volume of extracellular matrix around the cardiomyocytes. We could see consistent changes in composition of the extracellular matrix during the development of this latter type of reactive fibrosis. Initially we detected that the cellular form of fibronectin which contained the ED-A sequence became more abundant in the intracellular spaces, and later we observed an increased fluorescence of collagen type I. Ultimately after approximately 20 to 40 days there was a predominance of collagen but little fibronectin was shown in the fibrous tissue surrounding the myocytes.

The reparative fibrosis which we observed had a focal character and was most probably due to perioperative and postoperative ischemia of the myocardium which is unavoidable during the transplantation procedure. During the early periods after surgery a high content of fibronectin was observed in the granulation tissue, while a relatively low concentration of this glycoprotein was later observed in the permanent scars (not shown). Granulation tissue observed at 3 days after transplantation, indicate that local ischemia leading to tissue damage most probably occurred during the process of surgery or shortly afterwards, as a similar accumulation of fibronectin in the granulation tissue of myocardium was previously observed during myocardial infarction (9,11). The source of fibronectin in granulation tissue was basically twofold. First, it penetrated there, most probably from the bloodstream, and second, its production in situ was detected by monoclonal antibodies as well. The accumulation of fibronectin in granulation tissue most probably stopped later on and was replaced by scars containing predominantly type I collagen.

In contrast to reparative fibrosis, the source of fibronectin in the diffuse reactive fibrosis was most probably primarily due to local 'de novo' synthesis in the intracellular spaces. Consequently, this local cellular source in conjunction with additional plasma fibronectin transported to the intercellular spaces from elsewhere, then constitute the primary matrix for development of both types of fibrosis in the transplanted heart. A similar role of fibronectin was observed by others in healing wounds under experimental conditions (11,12,13,14,15).

Connective tissue and extracellular matrix are integral components of the myocardial wall in addition to cardiomyocytes and coronary vasculature. It has been shown that diffuse reactive cardiac fibrosis could be seen during the process of cardiac hypertrophy (16) and is considered to be a reaction to increased load imposed upon the myocardial tissue. In our experimental model we could also observe an enhanced expression of fibronectin and collagen which are normally synthesized in the fibroblasts. This occurred irrespective of the fact that the load of the heterotopically isotransplanted heart is considerably decreased as can be demonstrated both by direct measurements (17) and also inferred from the considerable atrophy of the myocytes (3). It may suggest that other factors besides overload may stimulate the release of various factors such as transforming growth factor ß (18,19,20), interleukin-1 (21,22), platelet derived growth factor (23) and interferon (24) which may then be responsible for enhanced production of various components of ECM by interstitial cells in the myocardium.

Connective tissue and some components of extracellular matrix are thought to be responsible for the passive visco-elastic properties of the cardiac chambers (25). Consequently any alteration in their relative volume or the character of their molecular network should alter the passive stress-strain properties of the cardiac chambers and may later interfere with the ventricular filling during diastole. In spite of the fact that the passive properties of the papillary muscle of long term transplants remained unaffected (2), recent evidence indicates that the passive pressure volume relationship of the left ventricles of the transplant show increase stiffness with time after transplantation (26). The above may limit the usefulness of this model for some quantitative hemodynamic studies which require longer term postoperative time periods.

Acknowledgements: The author's wish to thank Mrs. M. Masika for technical help and the Medical Research Council of Canada for financial support.

References

1. Korecky B, Masika M. Toxicologic Pathology 1991;18:541-546.
2. Korecky B, Rakusan K. In: Perspective of cardiovascular research (ed.) Aplert NR. 1983; New York, Raven Press.
3. Campbell SE, Korecky B, Rakusan K. Circ Res 1991;68:984-996.
4. Mosesson MW, Umfleet RA. J Biol Chem 1970;245:5728-5736.
5. Idelson GL, Muzykantov VR, Chekneva EE, Shnyra AA, Shekhonin BV Domogatsky SP. Collagen Related Res 1987;7:383-397.
6. Borsi L, Carnemolla B, Castellani P, Rosellini C, Vecchio D, Allemanni SE, Chang SE, Taylor-Papadimitriou J, Pande H, Zardi L. J Cell Biol 1987;104:595-600.
7. Carnemolla B, Borsi L, Zardi L, Owens RJ, Barelle FE. FEBS Letters 1987;215:269-273.
8. Chernousov MA, Faerman AI, Frid MG, Printseva O.Yu, Koteliansky VE. FEBS Letters 1987;217:124-128.
9. Shekhonin BV, Domogatsky SP, Idelson GL, Koteliansky VE. J Mol Cell Cardiol 1988;20:501-508.
10. Chapman D, Eghbali M. Cardiovasc Res 1990;24:578-583.
11. Shekhonin BV, Guriev SB, Irgaskev SB, Koteliansky VE. J Mol Cell Cardiol 1990;22:533-541.
12. Kurkinen M, Vaheri A, Roberts PJ, Stenman S. Lab Invest 1980;43:47-51.
13. Grinnel F, Billingham KE, Burgess L. Dermatol 1981;76:181-189.
14. Hoolung B, Clemmensen J, Junker P, Lyon H. Acta Micr Scand A 1982;90:159-165.

15. Repesh LA, Fitzgerald TJ, Furcht LT. J Hist Cytochem 1982;30:351-358.
16. Weber KT, Janicki JS, Shroff SG, Pick R, Abrahams C, Chen RM, Bashey RI. J Appl Cardiol 1988;3:37-46.
17. Korecky B, Masika M. Circ Res 1991;68:1174-1178.
18. Raghow R, Postlethwaite AE, Keski-Oja J, Moses HL, Kang AH. J Clin Invest 1987;79:1285-1288.
19. Fine A, Goldstein RH. J Biol Chem 1987;262:3897-3902.
20. Ignotz RA, Endo T, Massague Y. J Biol Chem 1987;262:6443-6446.
21. Kahari VM, Heino Y, Vuorio E. BBA 1987;929:142-147.
22. Postlethwaite AE, Raghow R, Stricklin GP, Poppleton R, Eyer JM, Kang AH. J Cell Biol 1988;106:311-318.
23. Narayanan AS, Page RC. J Biol Chem 1983;258:11694-11699.
24. Czaja MJ, Weiner FR, Eghbali M, Giambrone MA, Zern MA. J Biol Chem 1981;262:13348-13351.
25. Weber KT. J Am Coll Cardiol 1989;13:1637-1652.
26. Galinanes M, Hearse DJ. J Heart and Lung Transplant 1991;10:79-91.

www.ingramcontent.com/pod-product-compliance
Lightning Source LLC
Chambersburg PA
CBHW070744130926
51338CB00014B/23
* 9 7 8 0 7 9 2 3 2 0 5 2 4 *